# World Conference on Carbon (Carbon 2017)



Melbourne, Australia
23 - 28 July 2017

ISBN: 978-1-5108-7467-1

**Printed from e-media with permission by:**

Curran Associates, Inc.
57 Morehouse Lane
Red Hook, NY 12571

**Some format issues inherent in the e-media version may also appear in this print version.**



Printed by Curran Associates, Inc. (2019)

For permission requests, please contact CARBON 2017
at the address below.

CARBON 2017
C/O Professor Mark Biggs
Nottingham Trent University
Nottingham, United Kingdom

Phone: +44 (0)115 848 3840

mark.biggs@ntu.ac.uk

**Additional copies of this publication are available from:**

Curran Associates, Inc.
57 Morehouse Lane
Red Hook, NY 12571 USA
Phone: 845-758-0400
Fax: 845-758-2633
Email: curran@proceedings.com
Web: www.proceedings.com

# TABLE OF CONTENTS

**Nanoscale Environment-Interactive Carbon Engineering** .......... 1
*K. Kaneko*
**A Reversible Crosslinking Structures of Pan Fibers During Heating in Inert Atmosphere** .......... 3
*Liangxiao Zhou, Fei Huang, Weizhe Zhao, Yonggen Lu*
**Activated Carbon-Incorporated Cigarette Filter for Tar and Nicotine Removal** .......... 6
*Bunsita Kong-Gij, Sataporn Kunapisitkul, Pichaya Wadsungnoen, Pawan Boonyoung, Panudetch Treeweranuwat, Khanin Nueangnoraj*
**Activated Polymer of Intrinsic Microporosity (PIM-1) with Improved Electrochemical Performance for Lithium-Ion Batteries** .......... 9
*Yang Meng, Meng Qinghan, Cao Bing, Yu Yunhua*
**Application of Natural Carbon Fiber for Oil Spill Sorption** .......... 13
*Apiluck Eiad-Ua, Wachiraporn Gunpum, Nawin Viriya-Empikul, Kajornsak Faungnawakij*
**Arsenic(v) Electroadsortion: Effect of Surface Chemistry and Porosity of activated Carbon Electrodes** .......... 15
*Luis F. Chazaro-Ruiz, Rigoberto Santoyo-Cisneros, René Rangel-Méndez*
**Batteries Marry Air Pollutants: Recover $H_2S$ in Li-S Battery** .......... 17
*Chen Zhang, Donghai Liu, Wei Lv, Lijie Zhi, Quan-Hong Yang*
**Biomass-based Three-Dimensional Network of Carbon Nanosheets for Lithium-Ion Battery Anodes** .......... 19
*Shasha Guo, Yaxin Chen, Xiaohong Chen, Jisheng Zhou, Zhaokun Ma, Huaihe Song*
**Carbon Matrix for C/C Composite by Film Boiling Technique** .......... 21
*Yasushi Soneda, Norio Iwashita, Michimasa Uda, Hiroshi Yamauchi*
**Carbon Nanofiber Film as Binder-Free Anode Materials for Sodium-Ion Storage** .......... 24
*X. Guo, J. Zhou, H. Song, X. Chen*
**Carbon Nanoparticle Containing Hydrogel Nanocomposites with Enhanced IR Sensitivity** .......... 27
*Krisztina László, Barbara Berke, Lionel Porcar, Orsolya Czakkel*
**Carbon Spheres with Onion-like Structure As an Anode Material for Lithium-ion Batteries** .......... 30
*Bin Cao, Huaihe Song, Xiaohong Chen, Jisheng Zhou, Zhaokun Ma*
**Carbon Support Effects on Catalysts for Hydrogenation of Multifunctional Chemicals** .......... 33
*Radhika Rao, Raoul Blume, Thomas Hansen, Erika Fuentes, Kathleen Dreyer, David Hibbitts, Yves Chabal, Robert Schlögl, Jean-Philippe Tessonnier*
**$Co_{0.85}Se$ Nanosheets/Graphene Composite Film as Binder-free Anode for Lithium-ion Batteries** .......... 35
*G. Zhang, J. Zhou, X. Chen, H. Song*
**Confirmation of the Formation of Oxidative Debris During the Hummers-offeman Reaction in the Synthesis of Graphene Oxide** .......... 38
*G. Ramos-Fernández, J. Pérez-Galera, D. Domene López, I. Martin-Gullón*
**Construction of Integrated RGO/$MnFe_2O_4$/g-$C_3N_4$ Catalyst for Decomposition of Methylene Blue** .......... 41
*Jiangying Qu, Xiyue Peng, Chuang Geng, Feng Gao, Jieshan Qiu*
**Control of Pore Structures in MgO-templated Carbons for EDLCs** .......... 44
*Yuya Kado, Yasushi Soneda*
**Conversion of Kerosene to Carbon Nanotubes via Co-pyrolysis with Ferrocene** .......... 47
*Weerawut Chaiwat, Oukrit Thonganatakul, Parintorn Vaewhongs, Tawatchai Charinpanitkul, Komkrit Suttiponparnit*
**COR and OER of Several Carbon Materials in KOH Solution** .......... 50
*Taro Kinumoto, Makoto Eto, Kohei Ono, Miki Matsuoka, Tomoki Tsumura, Masahiro Toyoda*
**Defect Engineering of Graphene and Other 2D Materials** .......... 53
*Ruitao Lv, Mingxiang Hu, Feiyu Kang, Mauricio Terrones*
**Development of Carbon Spheres with Tailored Porosity, Structure and Functionality** .......... 56
*Mietek Jaroniec*
**Dielectric Behavior of Carbon Materials** .......... 58
*D. D. L. Chung*
**Doped Carbon Materials as Electrocatalysts for the Oxygen Reduction Reaction** .......... 60
*M. J. Nieto-Monge, J. C. Ruiz-Cornejo, D. Sebastián, C. Alegre, M. V. Martínez-Huerta, E. Pastor, M. J. Lázaro*
**Effect of Ball-milling on Supercapacitive Performance of Expanded Graphite** .......... 63
*Yue Dong, Su Zhang, Huaihe Song, Xiaohong Chen, Jisheng Zhou, Zhaokun Ma, Dianzeng Jia*
**Effect of Carbon Structure on Ru Catalyst for Ammonia Synthesis** .......... 65
*Masayasu Nishi, Saori Taira, Hideyuki Takagi*
**Electrochemical Oxidation of Carbon Xerogels and Correlation with Physicochemical Properties** .......... 68
*David Sebastián, Cinthia Alegre, María E. Gálvez, María J. Lázaro*
**Electrolytic Treatment Of Dry-jet Wet Spun Carbon Fibers** .......... 71
*Y. Ma, J. Liu, Y. Xue, J. Liang, X. Wang*
**Enhancement of Mechanical Properties of CFRTP Using Polymer Colloids** .......... 73
*Katsumasa Uematsu, Tetsuya Yamamoto, Toshihira Hrisawa, Yasuhiro Tanabe*
**Evaluation of Edges for Carbon Materials** .......... 75
*Kazuki Matsumura, Taro Kinumoto, Tomoki Tsumuta, Masahiro Toyoda*
**Fabrication and Anti-ablation Performance of W/Cu Composite Coatings by SAPS** .......... 77
*Ling-Jun Guo, Jian Peng*

**Fe-N Co-doped Graphene As Air-cathode Catalyst in Microbial Fuel Cells** ........ 79
*Dingling Wang, Zhaokun Ma, Yang'En Xie, Huaihe Song*
**Flexible Free-standing Graphene /Hollow Tin Dioxide Paper for Lithium-Ion Batteries** ........ 81
*Wanyuan Zhi, Tao Xu, Qinghan Meng, Bing Cao, Yunhua Yu*
**Formation and Catalytic Activity for Oxygen Reduction Reaction of Fullerenesoot- Derived Onion-like Carbons** ........ 85
*Machiko Takigami, Takuya Maie, Kumi Nariduka, Koji Takasu, Naokatsu Kannari, Takafumi Ishii, Jun-Ichi Ozaki*
**Graphene Doped Based Electrocatalyst for Fuel Cell Applications** ........ 87
*G. Lemes, J. M. Luque-Centeno, D. Sebastian, M. V. Martinez-Huerta, E. Pastor, M. J. Lazaro*
**Graphitization of Spin Coating Polymer** ........ 90
*Yasushi Soneda, Noriko Yoshizawa, Masaya Kodama*
**High Capacitance at High Scan Rate by Hybrid Graphene-carbon Xerogel Supercapacitors** ........ 93
*G. Ramos-Fernández, M. Canal-Rodriguez, A. Arenillas, J. A Menéndez, I. Martin-Gullón*
**High-strength and Modulus Carbon Nanotube Fibers** ........ 96
*Ok-Kyung Park, Hoikil Choi, Jaesang Yu, Jae Kwan Lee, Chong Rae Park Bon-Cheol Ku*
**Hydrogen Transfer Phenomenon on Carbon Materials with Different Edge Structures** ........ 99
*Daichi Umeda, Takaaki Togo, Yasuhiro Yamada, Satoshi Sato*
**Influence of Out-of-plane Compression Induced Damage Effects on the Mechanical Properties of C/C** ........ 101
*T. Behnisch, M. Thieme, R. Bohm, M. Gude*
**Interconnected Fe-Mn Fibrils on Activated Carbon to Increase As(V) Uptake** ........ 104
*J. J. Gutiérrez-Martínez, J. R. Rangel-Mendez*
**Investigation of Chemical States of Nitrogen Introduced to Fullerene-soot-derived Onion-like-carbons and Its Property Toward Oxygen Reduction Reaction** ........ 106
*Takafumi Ishii, Takuya Maie, Kumi Nariduka, Naokatsu Kannari, Machiko Takigami, Jun-Ichi Ozaki*
**Magnetic Carbon Nanosphere-Based Surface Molecularly Imprinted Polymers for Dibenzothiophene Adsorptive Removal** ........ 108
*Lei Qin, Weifeng Liu, Yongzhen Yang, Xuguang Liu, Lanqing Hu*
**Microwave Assisted Synthesis of Graphene-$Bi_8La_{10}O_{27}$-zeolite Nanocomposite with Efficient Photocatalytic Activity Towards Organic Dye Degradation** ........ 111
*Yonrapach Areerob, Won-Chun Oh*
**Microwave-Assisted Synthesis of C-doped $TiO_2$ and ZnO as Solar-cells Electrodes** ........ 113
*J. R. Rangel-Mendez, J. Matos, L. F. Cházaro-Ruiz, A. C. González-Castillo, G. Barrios-Yáñez*
**Modified Carbon Nanofiber Films as Interlayers for Lithium Sulfur Batteries** ........ 116
*G. Liang, X. Qin, Q. Li, M. Liu, B. Li, F. Kang*
**Narrow-diameter Distributed Single-walled Carbon Nanotubes Grown from Carbon Nanorings** ........ 119
*Tomohiro Tojo, Ryoji Inada, Yoji Sakurai*
**N-doped Carbon Quantum Dots Towards Electroluminescence** ........ 122
*Jingxia Zheng, Yuanfei Ding, Yanqin Miao, Yongzhen Yang, Xuguang Liu, Husheng Jia*
**N-doped Mesoporous Carbons Decorated with Pt Nanoparticles in Hydrotreatment Reactions** ........ 125
*Cristina Ruiz-García, Francisco Heras, Noelia Alonso-Morales, Luisa Calvo, Juan J. Rodríguez, Miguel A. Gilarranz*
**$NH_4FeF_3$/Carbon-Nanosheet Composite as a Potential Anode Material for Li Ion Storage** ........ 128
*Minhong Kong, Jisheng Zhou, Xiaohong Chen, Huaihe Song*
**Nitrogen Doped Carbon Aerogel - Graphene Composite Materials for Electrocatalysis** ........ 131
*Krisztina László, Balázs Nagy, Imre Bertóti, Miklós Mohai, István Bakos*
**Nitrogen-Doped Carbon Materials Analyzed by X-ray Photoelectron Spectroscopy** ........ 134
*Haruki Tanaka, Yasuhiro Yamada, Satoshi Sato, Shingo Kubo*
**Two-dimensional Nitrogen-doped Carbon Nanosheets Supported $NiSe_2$ Nanoparticles as Anode Material for Enhanced Sodium Storage** ........ 136
*Sitong Liu, Jisheng Zhou, Xiaohong Chen, Huaihe Song*
**Nitrogen-doped Carbon Nanotubes Derived from Metal–organic Framework As an Efficient ORR Electrocatalyst** ........ 138
*Rui Li, Xuzhen Wang, Yanfeng Dong, Lan Yang, Jieshan Qiu*
**Optimization of Molding Process for CFRTP Having High Thermal Stability** ........ 141
*Toshihira Irisawa, Ryo Hashimoto, Masahiro Arai, Yasuhiro Tanabe*
**Oxy-tetracycline Removal via Ozonation Enhanced with Synthesized Magnetic Carbon Nanoparticles** ........ 144
*Konrat Kerdnawee, Chompoopich Termvidchakorn, Karanick Menakanist, Napon Opasanon, Prakitr Srisuma, Nuttawat Suwattanapongtada, Noriaki Sano, Hajime Tamon, Tawatchai Charinpanitkul*
**Performance of Pd/graphene Oxide Catalysts: Influence of Reduction Degree** ........ 147
*Cristina Ruiz-García, Yu Lei, Francisco Heras, Ana Laura Elías, Mauricio Terrones, Miguel A. Gilarranz*
**Polyaniline-based Carbon Nanospheres As Anode Materials for Sodium-ion Batteries** ........ 150
*Chunli Zhou, Huaihe Song, Xiaohong Chen, Mengqiu Jia*
**Potential of Carbon Gels in Hydrogen Storage Systems** ........ 152
*Krisztina László, Orsolya Czakkel, Balázs Nagy, Emanuel Bahn, Peter Fouquet, Silvia Villar-Rodil, Juan M. D. Tascón*
**Preparation and Application of Flexible and Self-standing Nonwoven Carbon Nanofibers** ........ 155
*X. Yang, J. Liu, X. Wang, J. Liang, Y. Xue*
**Preparation and Investigation of SWCNT Coatings Suitable for Biosensor Applications** ........ 158
*Jurgis Barkauskas, Ieva Paklonskaite, Justina Gaidukevic, Arunas Ramanavicius, Lina Mikoliunaite, Povilas Genys, Jurate Jolanta Petroniene, Inga Morkvenaite–Vilkonciene*
**Probabilistically Based Defect Analysis and Structure-property-relations of CF** ........ 160
*D. S. Wolz, M. Thieme, M. Kirsten, M. Loffler, U. Muhle, B. Rellinghaus, H. Jager, M. Gude, E. Zschech, C. Cherif, R. Bohm*
**Processing and Characterization of Long Fiber Thermoplastic Technology-based Carbon Fiber/ABS Composites** ........ 162
*Donghwan Cho, Daekyun Hwang*

**Production of Cone-Shaped Graphitic Whisker in Scrolled Natural Polymer Film as "Microreactor"** ........ 164
*Yukie Saito, Valeriy Luchnicov*
**Progressive Fabrication of Graphene and their Potential Applications** ........ 166
*Won-Chun Oh*
**Radial Distribution Analysis of Carbon Fibers using Neutron Diffraction** ........ 168
*Kiminori Ono, Haruki Okuda, Masaru Nakada, Fumihiko Tanaka*
**Regioselective Catalysts for Repair of Graphene Structure from Graphene Oxide** ........ 170
*Jurgis Barkauskas, Justina Gaidukevic, Ieva Mališauskaite, Rasa Pauliukaite, Gediminas Niaura, Ieva Matulaitiene, Gvidas Astromskas, Virginijus Bukauskas, Romualdas Trusovas*
**Role of Porosity in the Hygroscopic Nature of Nanodiamonds** ........ 172
*Elda-Zoraida Piña-Salazar, Koki Urita, Takuya Hayashi, Eiji Osawa, Toshio Sakai, Katsumi Kaneko*
**Self-assembly of Nitrogen-doped Carbon Nanobelts via Catalytic Chemical Vapour Deposition** ........ 175
*Juan L. Fajardo-Díaz, Alejandro J. Cortés-López, Florentino López-Urías, Emilio Muñoz-Sandoval*
**Separation of Cone-Shaped Graphitic Whiskers by Density Gradient Centrifugation** ........ 179
*Yuko Goto, Yukie Saito*
**Effects of CNT Composite on Snow Sliding Property and Water Sliding Property on the Vibrated Surface of Silicone/ CNT Composite Sheets** ........ 182
*Kenji Yanagisawa, Mitsuki Tsuchiya*
**Solvothermal Synthesis of Surfactant Free Spherical Nickel Hydroxide/Graphene Oxide Composite for Supercapacitor Application** ........ 184
*A. A. Khaleed, A. Bello, J. K. Dangbegnon, B. W. Mwakikunga, N. Manyala*
**Super-high Rate Sodium-Ion Storage Performance of $FeSe_2$/Graphene** ........ 187
*Dan Li, Jisheng Zhou, Xiaohong Chen, Huaihe Song*
**Surface Modification of Carbon Fibers by Electroless Tungsten Deposition for Enhanced Electrochemical Stability of Lead-Carbon Interface** ........ 189
*Shu-Huei Hsieh, Yi-Ren Tzeng, Ya-Wun Jan*
**Surface Modification of Graphene and Graphene Oxides by Nitrogen Plasma: Determination of Chemical State Alterations by Quantitative X-ray Photoelectron Spectroscopy** ........ 191
*Krisztina László, Barbara Berke, Imre Bertóti, Miklós Mohai*
**Synthesis of Frost-like CuO Combined Graphene-$TiO_2$ for High Photocatalytic Performance** ........ 193
*Dinh Cung Tien Nguyen, Woo-Sik Kim, Kwang-Youn Cho, Won-Chun Oh*
**Synthesis of Graphene Based Materials and Their Fuel Cells Electrocatalyst Applications** ........ 195
*María Jesús Lázaro, Luis M. Rivera, César Montero, Gonzalo García, Elena Pastor*
**Tailoring Few-Layer Graphene for Supercapacitors Using a Hydrogen-Argon Mixed Plasma** ........ 198
*Emily A M Smith, Carol Crean, Robert Slade, Chris Spacie, Chris Stirling, David Watson*
**Thermodynamic Rationalization of Graphitization: the Critical Temperature Threshold** ........ 200
*Philippe Ouzilleau, Aïmen E. Gheribi, Patrice Chartrand*
**Thermophysical Properties of Carbon Fibre Reinforced Multilayered $(PyC–SiC)_n$ Matrix Composites** ........ 203
*Yan Jia, Kezhi Li*
**TiCo/Graphene Based Nanocomposites and Their Activity Towards Oxygen Reduction Reaction** ........ 206
*J. M. Luque-Centeno, M. V. Martinez-Huerta, G. Lemes, D. Sebastian, E. Pastor, M. J. Lazaro*
**Transverse Structure Control of Mesophase Pitch Based Carbon Fiber** ........ 209
*Zhulin Sun, Changlin Yang, Yonggen Lu*
**Two-Dimensional Carbon-Encapsulated $Fe_3O_4$ Nanosheets Grown Via Oriented Attachment Mechanism** ........ 212
*Jinyu Ning, Minhong Kong, Huaihe Song, Jisheng Zhou*
**Water Desalination with Graphene Oxide/Graphene Based Membranes** ........ 215
*Aaron Morelos-Gomez, Rodolfo Cruz-Silva, Hiroyuki Muramatsu, Josue Ortiz-Medina, Takumi Araki, Tomoyuki Fukuyo, Syogo Tejima, Kenji Takeuchi, Takuya Hayashi, Mauricio Terrones, Morinobu Endo*
**Author Index**

## Nanoscale Environment-Interactive Carbon Engineering

***Kaneko, K***

e-mail: kkaneko@shinshu-u.ac.jp
*Shinshu University, Nagano, Japan*

Graphene of large surface area of 2630 $m^2g^{-1}$ offers $\pi$-conjugated bi-surfaces, providing the strongest dispersion interaction per unit weight. Therefore, the graphene-based carbon materials have been widely applied to various surface sensitive technologies. As our human society must construct a better sustainably society, we need to get an essential understanding and tuning of the interfacial properties and graphene based carbons. This presentation will introduce recent studies on confinement of molecules and ions in the carbon nanopores, on designing the smart gate function, and on the aggregation control of nanoscale carbons.

**$^{18}O_2$ : $^{16}O_2$ separation.** Highly efficient separation of **$^{18}O_2$ and $^{16}O_2$** isotopic molecules. $^{18}O$ is indispensable to the positron emission tomography for cancer therapy. The concentration of $^{18}O$ is 0.2 %, being considerably high. However, the classical separation selectivity is the order of 1.003. Low temperature adsorption method of oxygen isotopes using nanoporous carbons gives more than **200 times higher** selectivity, as shown in Figure 1. Figure 1 shows the time change of the adsorption selectivity for $^{18}O_2/^{16}O_2$ with activated carbon fibers (ACF) of different pore widths (ACF5: 0. 7 nm, ACF10: 0.9 nm, ACF20: 1.1 nm) at 112 K in Introduction of the 1:1 mixed gas with [2/7-3.7] x 10-4 $s^{-1}$.The adsorption selectivity is quite high compared with the traditional method.

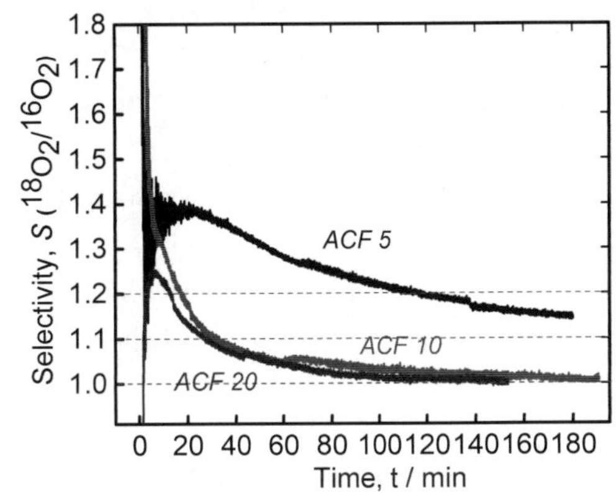


Figure 1. Adsorption selectivity for $^{18}O_2/^{16}O_2$ on three kinds of ACFS at 112 K.

**Non-Coulomibic Ion Structure.** Supercapacitors which support new sustainable technologies have requested structural understanding of ions confined in carbon nanopores. In particular, the ionic structure in subnanoscale pores has gathered a great attention. We trans fo rmed 1D synchrotron X-ray scattering information to the plausible 3D structure information through hybrid reverse Monte Carlo simulation, getting the component structure on cations and an ions. This study evidenced the **non-Coulombic accumulation** of ions having the same charge state in 0.7 nm-pores, supporting the super-ionic theory. Figure 2 shows the compensation mechanism of co-ions due to image charges produced in the carbon pore walls. This

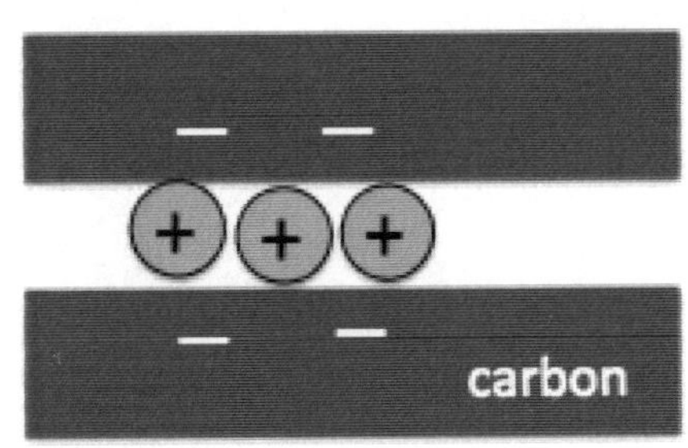


Figure 2. Mechanism of image charge-mediated Co-ions –accumulation in narrow carbon pores

unusual compensation mechanism works only for the narrow carbon pores whose width is close to the molecular ion size.

**Smart nanowindows.** Single wall carbon nanohorn (SWCNH) consists of a less-crystalline graphene wall and the nano windows of target size are donated in the graphene wall by regulated oxidation. $N_2$ molecules at 77 K can more rapidly penetrate the nanowindows of 4 nm than argon at 87 K. However, $N_2$ molecules are trapped by the $N_2$ quadrupole moment-local electric filed interaction enough long to accept further nitrogen molecules; the obtained surface area from nitrogen adsorption is larger than that from argon adsorption by 20 %, as shown in Figure 3. The new **quasi-wall effect** by nitrogen adsorbed at the nanowindows is evidenced. Also the **concerted motion of the nanowindow rim** gives rise to rapid penetration of the molecules larger than the geometrical nanowindow size. The information provides a precious insight to a new generation air separation membrane.

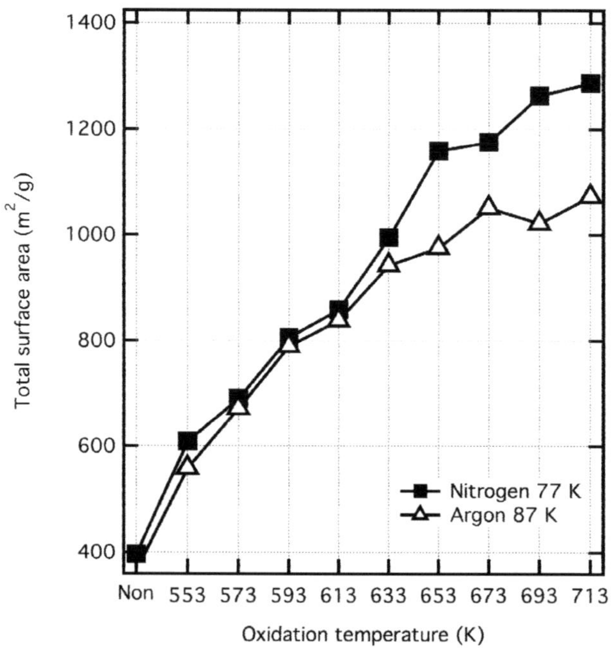


Figure 3. Total surface area difference between adsorption of $N_2$ and Ar for SWCNH heated at different temperatures

**SWCNT dispersant.** Carbon materials are hydrophobic and the interaction of carbons with water is quite unique. Single wall carbon nanotubes(SWCNTs) are associated with each other to form the bundle structure. An efficient dispersion of the bundles in water has been requested to apply SWCNT to various technologies. The surfactants are widely used for dispersion of the SWCNT bundles; the surfactants strongly bound on SWCNT deteriorate the excellent properties of SWCNT. Newly developed Zn/Al sol-gel dispersant can disperse the SWCNT bundles to produce a highly stable SWCNT ink, which leads to transparent conducting film and stretchable electrodes.

# A reversible crosslinking structures of PAN fibers during heating in inert atmosphere

Liangxiao Zhou, Fei Huang, Weizhe Zhao, Yonggen Lu*,

State Key Laboratory for Modification of Chemical Fibers and Polymer Materials, College of Materials Science and Engineering, Donghua University, Shanghai, 201620, China
*yglu@dhu.edu.cn

## ABSTRACT

It is found that chemical crosslinking take place between nitrile groups of adjacent polyacrylonitrile molecules when heated at temperature between 170-230 °C in inert atmosphere, which is proved by its insolubility in concentrated sulfuric acid. However, after heat treatment in an inert atmosphere above 230 °Cfor one hour, the derived fibers become resolvable again in sulfuric acid, meaning the formed crosslinking structure, probably -C=N-N=C- bonds, is reversible. Oxygen in the atmosphere inhibit such a kind of crosslinking even oxidation is necessary for a thermostable and irreversible structure.

**Key words:** polyacrylonitrile fibers, crosslinking, nitrile groups, heating, inert atmosphere

## INTRODUCTION

It is well known that stabilization is necessary before carbonization for carbon fiber manufacture. For polyacrylonitrile(PAN) based carbon fiber, the stabilization comes from the cyclization and oxidation crosslinking by heating in air. It have been considered that no crosslinking forming without oxidation [1)]. Nevertheless, in our previous work [2)], we found that crosslinking formed in PAN fibers during gamma ray irradiation in vacuum. Therefore, it is doubted if a crosslinking structure has ever taken place in the process of heating. In this work, we tried to investigate such a crosslinking structure during heating PAN fiber in vacuum compared to in air at a series of temperatures.

## MATERIAL AND METHODS

PAN precursor fibers (wet-spun from acrylonitrile/methyl methacrylate copolymer with weight ratio 99/1, and having 3000 filaments/tow, 12.5 μm average diameter) were supplied by a domestic corporation. These fibers were heated at 170, 190, 210, 230 and 250 °C for 2 hours in vacuum and air respectively. The gel fractions were analyzed by using sulfuric acid as the previous work [2)].

## RESULTS AND DISCUSSION

*Gel fractions.* Fig. 1 shows the generated gel fraction of the PAN fibers when heated at different temperatures for 2 hours. It is noted that PAN fibers heated in vacuum produced gels at temperatures from 170 to 210 °C, and disappeared when temperature rose to 230 °C. In air, gel did not emerge at lower temperature until 210 °C, however, the gel fraction increased when heating above 210 °C. It is demonstrated that crosslinking do have taken place by heating without oxidation, however, the crosslinking disappeared when heated to higher temperature. It is considered that the crosslinking structure without air was attributed to -C=N-N=C- bond formation between nitrile groups when heating in vacuum and oxygen inhabited such a kind of structure generating.

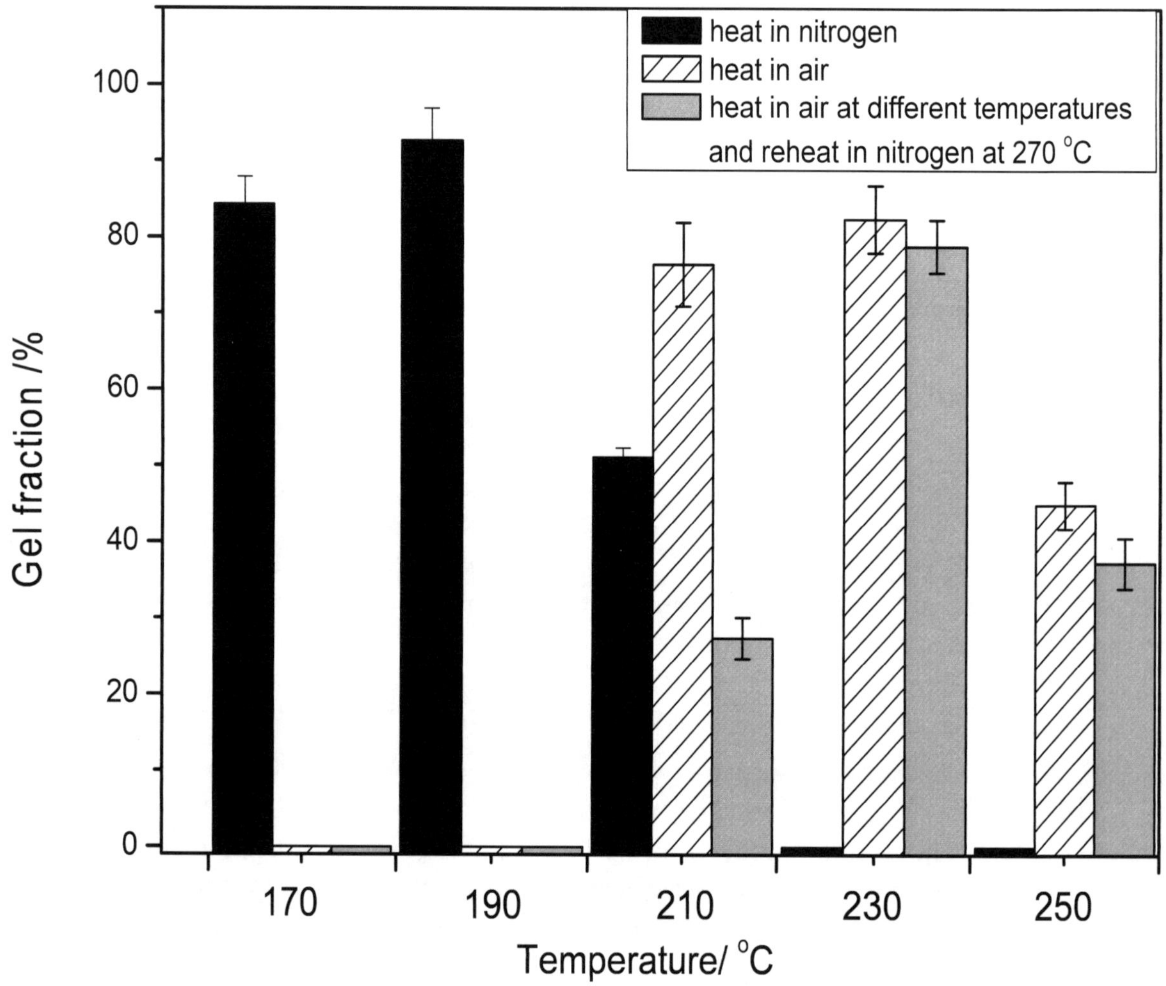


**Fig. 1. Gel fractions of PAN heated in air and nitrogen at different temperatures**

In order to investigate if the nitrile crosslinking took place when heating in air, the fiber samples heated in air was reheated at 270 °C in vacuum. It is found that the gel fraction decreased for the sample heated at 210 °C, and little differences of gel fraction before and after reheating were observed when the samples were heated at higher temperatures. It is noted that the nitrile crosslinking is reversible even it is chemically stable compared to oxidation crosslinking structure.

## SUMMARY

Nitrile crosslinking takes place at lower temperature (170~210 °C) and gradually disappears at higher temperature (above 230 °C) in PAN fibers during heating without oxygen. The nitrile crosslinking is inhibited by oxygen and replaced by oxidation crosslinking when PAN fibers are heated in air.

## REFERENCES

1. Gupta, A., and Harrison, I. R. (1996). "New aspects in the oxidative stabilization of pan-based carbon fibers". *Carbon*, Vol.34(11), p.1427-14450

2. Zhou, L., Lu, Y., Zhao, W., Yang, C., and Jiang, J. (2016). "Effects of gamma ray irradiation on poly(acrylonitrile-co-methyl acrylate) fibers". *Polymer Degradation and Stability*, Vol.128, p.149-157

# Activated Carbon-Incorporated Cigarette Filter for Tar and Nicotine Removal

Bunsita Kong-gij, Sataporn Kunapisitkul, Pichaya Wadsungnoen, Pawan Boonyoung, Panudetch Treeweranuwat, Khanin Nueangnoraj*

School of Bio-Chemical Engineering and Technology, Sirindhorn International Institute of Technology, Thammasat University; Thailand.
* khanin@siit.tu.ac.th

## INTRODUCTION

Cigarette smoke is one of the world environmental issue, causing several health problems to both of the smokers and non-smokers. Tar and nicotine are the most important matters in the smoke, which directly affects the respiratory system, such as lung cancer and emphysema. Adding a filter into the cigarette can partly screen out such the harmful substances. Additives, such as zeolites and activated carbon can be added into the filter to improve the adsorption ability. However, not only the high specific surface area, which lead to the enhancement of physisorption, the surface chemistry of solid additive is also important to employ the chemisorption. This work aims to develop the activated carbon that can effectively remove tar and nicotine from the cigarette mainstream smoke. Activated carbon was derived from hemp (*Cannabis sativa*), using chemical activation with phosphoric acid. Process parameters such as activation temperature and impregnation ratio will be investigated. The resultant activated carbons will be fully characterized to understand the effect of pore structure and surface chemistry of the carbons on the adsorption ability of tar and nicotine from cigarette mainstream smoke.

## MATERIALS AND METHODS

### Material

Hemp stems were supplied by Thailand Tobacco Monopoly. Phosphoric acid (85%w/w) was used as an activating agent. SMS (Red) cigarette was used as the reference.

### Activated Carbon Preparation

Hemp stems were dried and grinded into a powder before impregnating with phosphoric acid at room temperature and then dried at 105°C for 48 hours in an oven. The impregnation ratio (hemp powder to phosphoric acid) of 1 and 2 were studied. The impregnated hemp was carbonized under $N_2$ flow (200cm$^3$STP/min) for 2 hours in a tubular furnace. The activation temperature ranges from 350 to 550°C with a heating rate of 10°C/min. After that, the samples were washed and dried at 105°C overnight. The ACs was labeled in Cn-m, while n is impregnation ratio of phosphoric acid and m is activation temperature. For example, C1-450 is activated carbon with impregnation ratio of 1 and activated at 450°C.

### Characterization

The surface chemistry on ACs was determined by Fourier Transform Infrared Spectroscopy (FTIR) with transmission technique using KBr pellet. The porous structure and pore size distribution were characterized by nitrogen sorption measurement at -196°C.

**Adsorption Ability**

The smoking machine was connected with a pump to draw the smoke through the filter. Tar was determined by the weight of brown substance on the cotton pad after burning cigarettes. While the cigarette was burning, mainstream smoke was trapped with Tip Extraction Standard (TES). The solution was characterized to estimate the amount of nicotine by gas chromatography analysis with flame ionization detector (GC/FID).

## RESULTS AND DISCUSSION

FTIR reveals that without activating agent, the carbon sample (C-450) has a similar functional groups to those in cellulose (the in-plane bending vibrations of C-H at 870 $cm^{-1}$). By adding the agent, such functional groups are converted to C-O stretching (in alcohol) and C=O stretching (in ketones, aldehydes and carboxylic acid). C1-350 contains a lot of polar functionalities, such as C=O, C-O and O-H. As the activation temperature increases, the intensity of the above functional groups was decreased. In addition C=C aromatic stretching vibrations at 1600 $cm^{-1}$ was also observed and it decreases at higher temperature.

**Table 1.** Specific surface area, tar and nicotine reduction on different ACs

| | $S_{BET}$ ($m^2/g$) | $\%R_{tar}$ | $\%R_{nicotine}$ |
|---|---|---|---|
| C-450 | 164 | 78.29 | 85.95 |
| C1-350 | 1160 | 65.89 | 93.20 |
| C1-450 | 815 | 50.00 | 48.09 |
| C1-550 | 1140 | 55.56 | 72.63 |
| C2-350 | 1880 | 67.18 | 85.56 |
| C2-450 | 2280 | 59.68 | 83.13 |
| C2-550 | 1400 | 52.97 | 22.29 |

C-450 has the lowest specific surface area (**Table 1**) and it greatly increases with the addition of phosphoric acid. Increasing the amount of phosphoric acid (or impregnation ratio) can further lead to the porosity development and enhance the specific surface area. However, only the effect of activation temperature was significant on the functional groups, not on specific surface area.

According to **Table 1**, increasing the impregnation ratio will decrease tar adsorption ability. This could be a result from the development of oxygen-containing functional groups that may not adsorb the non-polar polycyclic aromatics hydrocarbon (PAHs) such as tar[1)]. Therefore, C-450 shows the highest tar reduction. Unlike tar, nicotine acts as the polar molecules, which can adsorb effectively on C-350 that contains a lot of polar functionalities. It has been reported that the present of C=O stretching of carboxylic groups could facilitate the adsorption of nicotine[2)]. Therefore, the samples prepared at 350-450°C, which has C=O groups, can reduce nicotine more than 80%.

## CONCLUSIONS

The activated carbon was successfully prepared from hemp stem. Without the addition of phosphoric acid, it can effectively adsorb tar. By adding phosphoric acid, oxygen functional groups were development, which decrease the non-polar tar adsorption ability. However, this can enhance nicotine adsorption, especially C=O in carboxylic groups.

**REFERENCES**

1. Jia, B., Jia, X., Hyun Kim, K., Ji Pu, Z., Kang, M. S., Ok Jeon, C. (2017). "Evolutionary, computational, and biochemical studies of the salicylaldehyde dehydrogenases in the naphthalene degradation pathway", *Sci Rep 7*, 43489.

2. Chen, Z., Zhang, L., Tang, Y., Jia, Z. (2006). "Adsorption of nicotine and tar from the mainstream smoke of cigarettes by oxidized carbon nanotubes", *Applied Surface Science 252*(8), 2933-2937.

# ACTIVATED POLYMER OF INTRINSIC MICROPOROSITY (PIM-1) WITH IMPROVED ELECTROCHEMICAL PERFORMANCE FOR LITHIUM-ION BATTERIES

**Yang Meng[1], Meng Qinghan[1]*, Cao Bing[1], Yu Yunhua[2]**

[1]*(College of Materials Science and Engineering, Beijing University of Chemical Technology, Beijing 100029, China)*

[2]*(State Key Laboratory of Organic-inorganic Composites, Beijing University of Chemical Technology, Beijing 100029, China)*

*(corresponding author: qhmeng@mail.buct.edu.cn)*

## INTRODUCTION

With the rapid development of industrialization, the application of portable devices is becoming more and more widely. Followed by it, the demand for lithium-ion batteries (LIBs) of high performance also soar[1]. In order to obtain LIBs with high performance, excellent anode material is indispensable. The intrinsic microporous structure of PIM-1 makes it have free volume as high as 26%, and the specific surface area of it is also more than 500 $m^2 \cdot g^{-1}$ [2, 3]. And the porous structure of electrode materials can make the point of intercalated-$Li^+$ increase. The microporous can retain $Li^+$ and its compounds effectively, while the mesoporous can provide a large number of channels to transport $Li^+$, and it also confer higher ionic conductivity of the electrode[4]. In this work, PIM-1 was obtained by organic synthesis, and $CO_2$ activation was used to optimize the porous structure of PIM-1. Moreover, we used it as a anode material for LIBs.

## METHODOLOGY

5,5',6,6'-Tetrahydroxy-3,3,3',3'-tetramethyl-1,1'-spirobisindane(TTSBI) was purified by recrystallization in methanol and 2,3,5,6-tetrafluoroterephthalonitrile (TFTPN) was purified by vacuum-distillation. By a nucleophilic substitution reaction, PIM-1 was synthesized via TTSBI and TFTPN. Under the protection of $N_2$, the appropriate amount of TTSBI and TFTPN (molar ratio was 1: 1) was added to a three-necked flask, adding proper amount of DMF as solvent and $K_2CO_3$ as the catalyst. The above process was carried out in a constant temperature water bath at 60°C, and the reaction continued for 20h until the product becomes bright yellow and more viscous. Then the obtained product was removed, and washed by 0.1wt% HCl, deionized water and methanol several times. Finally put it in a vacuum oven with the temperature of 70°C, then PIM-1 was obtained. The dried PIM-1 was carbonized at 1000°C for 1h with temperature rising rate of 5°C/min under $N_2$ flow and activated by $CO_2$ flow for 4h, then the resulting material was denoted as PIM-1-A. For comparison, we also prepared PIM-1-C by similar method by carbonizing at 1000 °C for 5h with temperature rising rate of 5°C/min under $N_2$ flow.

The electrode materials were fabricated by mixing the active material (PIM-1-C or PIM-1-C), acetylene black and polyvinylidene fluoride (PVDF) with N-methyl pyrrolidone (NMP) as solvent with the weight ratio of 80:10:10. After grinding, the anode electrode was coated onto a piece of a copper foil and dried in a vacuum oven with the temperature of 80°C. In order to measure the performance of each electrode materials, the anodes were assembled as half-cells using CR2025 coin-cells separately. In this process, lithium foil acted as the counter/reference electrode, and 1M $LiPF_6$ in ethylene carbonate/dimethyl carbonate/ ethyl methyl carbonate (1:1:1 v/v/v) acted as the electrolyte solution, and propylene film (Celgard 2400) acted as the separator. Before tests, the half-cells were stored for 24h.

The galvanostatic charge-discharge (GCD) figures were obtained by a battery testing system (LAND CT2001A) at various current densities in the potential range of 0-3.0 V vs. $Li/Li^+$.

**RESULTS AND DISCUSSION**

Figure 1 shows the $N_2$ adsorption-desorption isotherms and pore size distribution curves of PIM-1-C and PIM-1-A. According to IUPAC classification, the $N_2$ adsorption-desorption isotherms of PIM-1-C is Type ?, which indicates the presence of a small amount of microporous. And the $N_2$ adsorption-desorption isotherms of PIM-1-C is mainly a combination of Type I and IV, which is of high adsorption under low relative pressure. And when the relative pressure range from 0.4 to 1.0, it appears a small hysteresis loop. In Figure 1b, the pore size distribution curve can clearly present that micropores and mesopores of PIM-1-A are more than PIM-1-A. And the porosity data in Table 1 shows that $CO_2$ activation increases the proportion of micropores of PIM-1-A, then improves its specific surface area, reduces average pore diameter.

Figure 2 shows the electrochemical properties of PIM-1-C and PIM-1-A. Figure 2a and Figures 2b show the GCD curves of PIM-1-C and PIM-1-A in a potential range of 0-3 V with different current densities. The GCD curves of PIM-1-C increase quickly under high potential which indicates that the intercalated-$Li^+$ points is less before activation, so that embedded quantity of $Li^+$ reaches saturated. Respectively, the discharge specific capacity of PIM-1-C and PIM-1-A is 230.5 $mAh{\cdot}g^{-1}$ and 1158.4 $mAh{\cdot}g^{-1}$ with a current density of 0.1 $A{\cdot}g^{-1}$, and 34.7 $mAh{\cdot}g^{-1}$ and 283.9 $mAh{\cdot}g^{-1}$ with a current density of 2 $A{\cdot}g^{-1}$. Figure 2c shows the rate capabilities of PIM-1-C and PIM-1-A, which indicates that $CO_2$ activation can make PIM-1 improve the capacity as well as keep the cycle stability. Besides, Figure 2d shows the first 100 cycles of PIM-1-C and PIM-1-A, which indicates that cycle stability of PIM-1-C is better than PIM-1-A.The specific capacities of PIM-1-A decline rapidly in the former 20 cycles, and tends to be stable after 39th cycle. The outstanding electrochemical performance of PIM-1-A is attributed to $CO_2$ activation which makes electrode materials have a large amount of micropores and a high specific surface area.


## ACKNOWLEDGMENT

This study is supported by National Nature Science Foundation of China (51372011).


## REFERENCES


1. Harks, P.P.R.M.L., et al. (2015). Journal of Power Sources, 288 (2015), 92-105.
2. Budd, P.M., et al.(2005). Journal of Materials Chemistry, 15(20), 1977-1986.
3. Mckeown, N.B., et al.(2006). Chemical Society Reviews, 35(8), 675-683.
4. Liu, N.P., et al. (2013). Wuli Huaxue Xuebao, 29(5), 966-972.


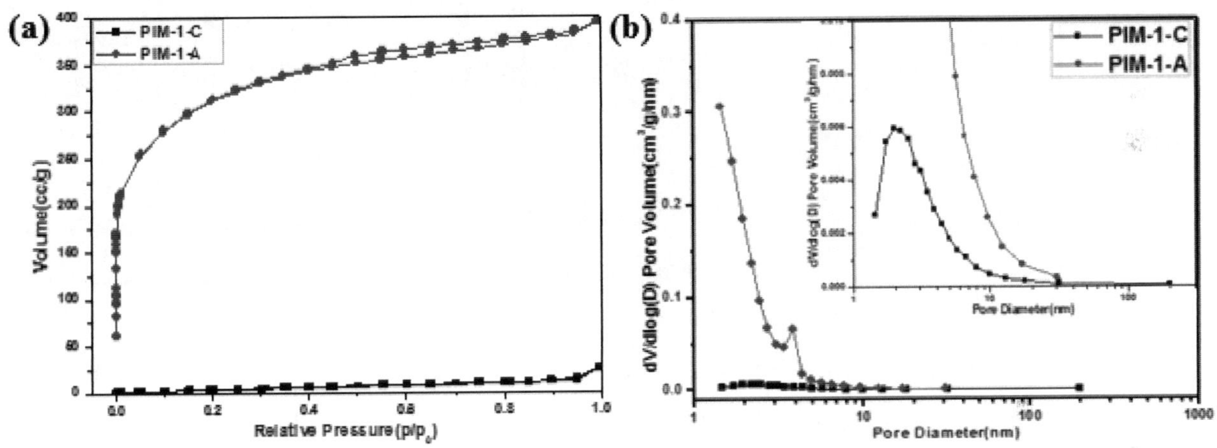


Fig. 1 (a) The $N_2$ adsorption-desorption isotherms of PIM-1-C and PIM-1-A; (b) The pore size distribution of PIM-1-C and PIM-1-A

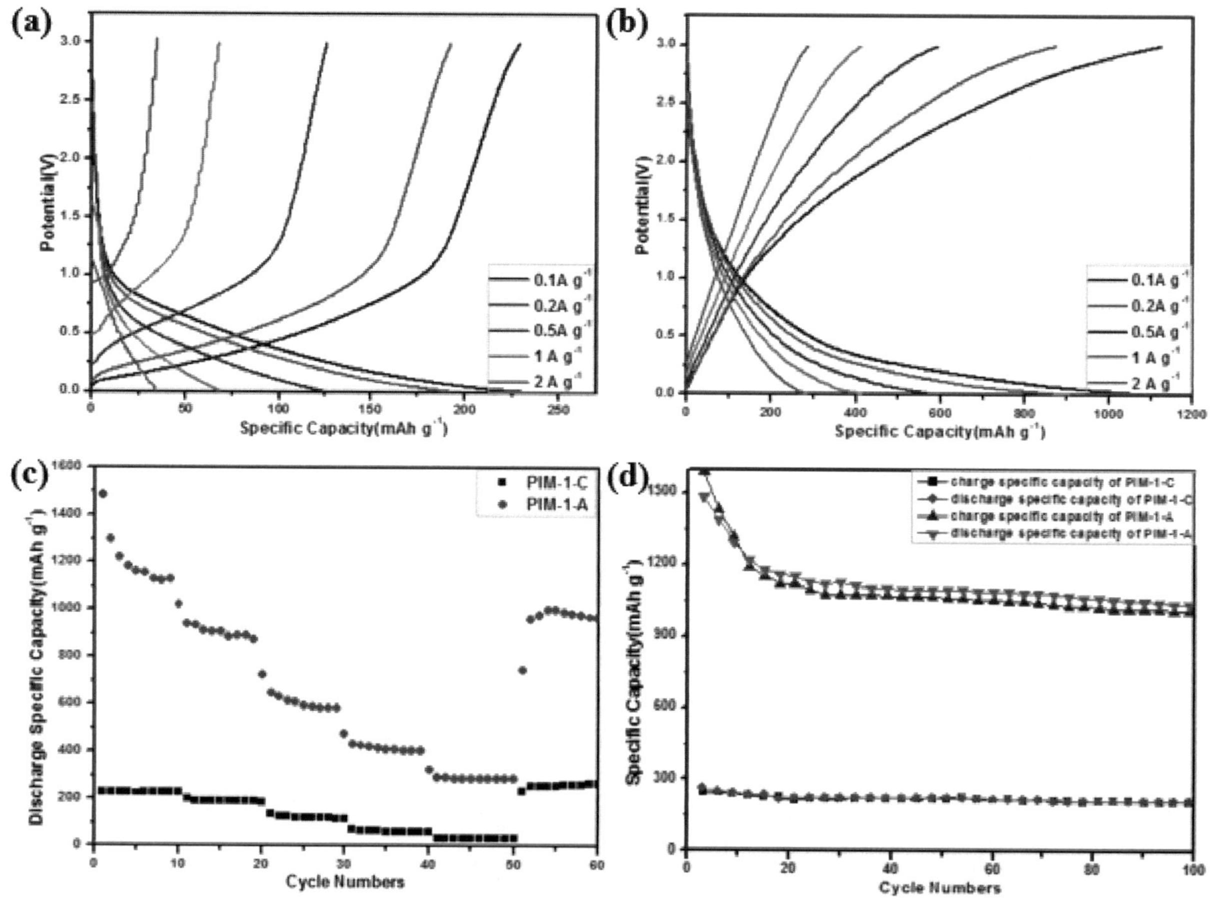


Fig. 2 (a) GCD curves of PIM-1-C at different current densities; (b) GCD curves of PIM-1-A at different current densities; (c) )The rate capabilities of PIM-1-C and PIM-1-A; (d) The cycle life curves of PIM-1-C and PIM-1-A with a current density of 0.1 $A \cdot g^{-1}$

Table 1 Specific surface area and porous structure analysis of PIM-1-C and PIM-1-A samples

| **Sample** | **$S_{BET}$ ($m^2 \cdot g^{-1}$)** | **$S_{mic}$ ($m^2 \cdot g^{-1}$)** | **$V_{tot}$ ($cm^3 \cdot g^{-1}$)** | **$V_{mic}$ ($cm^3 \cdot g^{-1}$)** | **$D_{ap}$ (nm)** |
|---|---|---|---|---|---|
| PIM-1-C | 133.77 | 77.25 | 0.124 | 0.092 | 2.671 |
| PIM-1-A | 965.3 | 832.9 | 0.610 | 0.457 | 2.528 |

# Application of Natural Carbon Fiber for Oil Spill Sorption

Apiluck Eiad-ua[1*], Wachiraporn Gunpum[1], Nawin Viriya-empikul[2], and Kajornsak Faungnawakij[3]

[1]College of Nanotechnology, King Mongkut's Institute of Technology Ladkrabang, Bangkok, Thailand
[2] The Joint Graduate School of Energy and Environmeatal, King Mongkut's University of Technology Thonburi, Bangkok, Thailand
[3]National Nanotechnology Center, National Science and Technology Development Agency, Thailand Science Park, Pathumthani, Thailand
*apiluck.ei@kmitl.ac.th

## INTRODUCTION

Oil spill is the one of the environmental concern nowadays. Pollution by petroleum oils affects sea life, economy and tourism because the coating properties of these materials.
The methods commonly used to remove oil involve oil booms, dispersants, skimmers, sorbents and etc. The main limitation of some of these techniques are their high cost and inefficiency trace level adsorption. Cattail flower (CF) have hydrophobicity (hydrophobic surface i.e., surfaces with contact angles greater than 90°[1]), non-polar surface and the roughness of surfaces at micro/nanometer length scale [2, 3]. The hydrothermal carbonization (HTC) is a thermochemical conversion process of biomass that yields a solid product, known as hydrochar. The advantages of HTC are naturalness, expediently, low temperature, low cost, no pollution and eco-friendly [4]. Our study show that carbon fiber via hydrothermal carbonization have high sorption capacity and high sorption rate for oil.

## MATERIALS AND METHODS

### Material

Cattail Flower (CF) was collected from stagnant water areas and beside the railway in Ladkrabang, Bangkok, Thailand. Then CF was dried in vacuum oven at 80°C for 12h to remove water and consecutively weighed until no further weight loss. Initially, 30 g of dried Cattail Flower (CF) and 60 ml of de-ionized water was mixed and transferred to glass liner and then put into the stainless steel reactor. The hydrothermal reaction was performed at the reaction temperature of 140, 160, 180, 200 and 220°C for 2, 4, 8, 12 and 24 h. After the reaction, the autoclave reactor was quenched in an ice bath to stop the reaction. The product was placed into a vacuum oven at 80°C for 12 h to completely remove moisture to obtained carbon product. Used oil, crude oil and slop oil were used in the oil adsorption test with porous carbon materials obtained from HTC.

## RESULTS AND DISCUSSION

The hydrothermal process at 180°C for 8h found that the highest oil adsorption about 18.93g/1g of sample (porous carbon material) and 15.18g/raw material (Cattail flower) because of the structure of CF sample was broken as small fiber (the area in contact with oil increased) and hydrophobic functional groups (carbonyl and benzene groups) increased. The CF (Pure) was the oil adsorption lower than hydrothermal of 180°C 8h because of without applying hydrothermal process (the structure could not improve) resulted in the structure of CF could not broken as small fiber structure (the area in contact with oil less than hydrothermal of 180°C 8h) and hydrophobic functional groups (carbonyl and benzene groups) also less than hydrothermal of 180°C 8h. However, the roughness (nano carbon formation as spheres) on the surface of CF sample increased and functional groups were eliminated. Only almost pure carbon (the increasing of non-polar) was investigated in the final product resulted in the hydrophobicity increased.

**Fig.1.** Oil Adsorption/1g of sample of hydrothermal process of CF at (A) different temperature of 140°C, 160°C, 180°C, 200°C and 220°C for 8h and (B) different time of 2h, 4h, 8h, 12h and 24h for 180°C.

## CONCLUSIONS

Cattail Flower (CF) was converted to porous carbon material by hydrothermal carbonization (HTC) improved the hydrophobicity, surface area property, functional group and morphology of CF for oil adsorption. The hydrothermal process at 180°C for 8h was the highest oil adsorption about 18.93g/1g of sample (porous carbon material) and 15.18g/raw material (Cattail flower). The hydrothermal process could break the CF sample to short molecule and split some part of fiber cluster to small fiber resulted in the area in contact with oil increased and hydrophobic functional groups also increased.

## REFERENCES

1. Barry Arkles. (1992). "Hydrophobicity, Hydrophilicity and Silanes, Paint & Coatings Industrymagazine", Vol.105.
2. Herminghaus, S. (2000). "Roughness-induced non-wetting", Europhysics Let., Vol.52, p.165-170.
3. Darhuber, A., Trojan, S.M. and Miler, S. (2000). "Morphology of liquid microstructure on chemicallypatterned surfaces", J. App Phy., Vol.87, p.7768-7775.
4. Sevilla M., Macia-Agullo JA., Fuertes A. (2013). "Biomass and Bioenergy", Vol.35, p.3152-3159.

# Arsenic(v) electroadsortion: Effect of Surface Chemistry and Porosity of activated Carbon Electrodes

Luis F. Chazaro-Ruiz[1], Rigoberto Santoyo-Cisneros[1], René Rangel-Méndez[1*]

[1]División de Ciencias Ambientales, Instituto Potosino de Investigación Científica y Tecnológica IPICYT, San Luis Potosí, S.L.P., Mexico
*luis.chazaro@ipicyt.edu.mx

## INTRODUCTION

The presence of arsenic in natural waters is a global problem, due to its toxic effects; it has been classified as a priority pollutant in different countries. Typical chronic effects of arsenic exposure include skin diseases as well as cardiovascular, neurological, renal and respiratory disorders. The elimination of arsenic from water has attracted much attention from the scientific community allowing the development of technologies to remove arsenic from water.
In spite of their simplicity, the adsorption methods are generally inefficient in reducing the arsenic concentration to acceptable levels (below 10 ppb) in drinking water, since they require highly operational conditions such as pH and temperature control, low selectivity, regeneration capacity, etc. This problem has prompted research into alternative treatments that can be used for the removal of arsenic at low concentrations. An example of such processes is the separation technologies driven by the action of an electric field, such as electroadsorption. The yield in the electroadsorption process is strongly dependent on the physical and chemical properties of the electrode materials[1], including, pore arrangement, specific surface area, and pore size distribution. It is well known that carbon materials have preferable characteristics such as high surface area and high conductivity. In this work we will present an study of the influence of the electrical conductivity, specific surface area, surface charge of a commercial granular activated carbon on the electroadsorption of As(V).

## MATERIALS AND METHODS

The activated carbon material used was granular activated carbon F400 (CA F400), of microporous and mesoporous nature. The physicochemical and electrochemical characterization of the material was made by nitrogen physisorption, potentiometric titrations and charge distribution (Bohem titration), and cyclic voltammetry (performed in $H_2SO_4$ 1 M for the specific capacitance determination, and electrochemical impedance spectroscopy. The electrodes were prepared by packing the CA F400 in stainless steel meshes as electric current collectors.

*Electrochemical system for electroadsorption experiments.* The experimental arrangement consisted of a 3-electrode system, placed in a 250 mL glass cell. The reference system (ER) used was an Ag/AgCl/KCl (sat) electrode, a stainless steel mesh as an auxiliary electrode (EA) and the activated carbon packed in the steel mesh as a working electrode (WE). The electrodes were exposed to the arsenic solution for approximately 12 hours without the application of an electrical potential, so that only the physical adsorption process was carried out. In the second phase, two potential of 0.4 and 0.8 V (*vs* Ag/AgCl/(KCl sat)) were applied to evaluate electroadsorption of arsenic. Aliquots of 1 mL from the electrochemical cell were periodically taken to determine the As(V) concentration by Inductively Coupled Plasma Mass Spectrometry (ICP-MS).

## RESULTS AND DISCUSSION

Figure 2 shows the adsorption kinetics of arsenic(V) divided in three stages: Stage I without voltage application in the cell in which the adsorption process occurs, reaching a maximum removal of 29% of the total concentration. The initial concentration was 1.7869 mg/g As(V). Stage II, electroadsorption, in which a voltage of 0.4 V was applied causing an additional 5% removal to the adsorption stage. In Stage III an additional removal of10% of the total concentration was achieved.

000

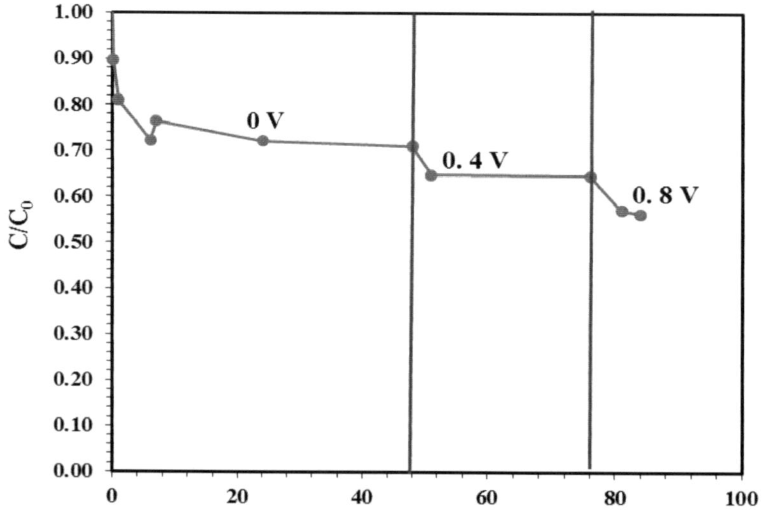


**Fig. 1**. Kinetics of adsorption and electroadsorption of As(V) in a three electrode cell. The applied voltages were refereed to Ag/AgCl/(sat KCl).

## CONCLUSIONS

Granular activated carbon F400 is a material that can be used as electrode material in electroassortion processes of arsenic in water, due to its natural ability to remove arsenic and its high specific capacitance. The application of anode potentials, 0.4 and 0.8 V on the material, increased its arsenic adsorption capacity by 15%, which is attributed to the charge of the double electric layer.


## ACKNOWLEDGMENTS

We thank for technical support from Elizabeth Cortez, Dulce Partida, Anely Fernandez. Also, thanks to LAMBAMA for the analysis and María del Carmen Rocha for technical support and CONACYT for financial support through project No. 789.


## REFERENCES


1. Noked, M., Soffer, A., & Arubach, D. (2011). "The electrochemistry of activated carbonaceous materials: Past, present, and future", *Journal of Solid State Electrochemistry*, *15*(7–8), 1563–1578.

2. Rasines, G., Lavela, P., Macías, C., Haro, M., Ania, C. O., & Tirado, J. L. (2012). "Electrochemical response of carbon aerogel electrodes in saline water", *Journal of Electroanalytical Chemistry*, *671*, 92–98.

# Batteries marry air pollutants: recover $H_2S$ in Li-S battery


Chen Zhang[1], Donghai Liu[2], Wei Lv[3], Lijie Zhi[4], Quan-Hong Yang[2, 3, 5*]

[1] School of Marine Science and Technology, Tianjin University, Tianjin, 300072, China
[2] School of Chemical Engineering and Technology, Tianjin University, Tianjin, 300350, China
[3]Graduate School at Shenzhen, Tsinghua University, Shenzhen, 518055, China
[4] National Center for Nanoscience and Technology, Beijing, 100150, China
[5]Tsinghua-Berkeley Shenzhen Institute (TBSI), Tsinghua University, Shenzhen, 518055, China
*qhyangcn@tju.edu.cn


## INTRODUCTION

Air pollution has a negative impact on the human health and quality of life, and environmental protection combined with advanced clean energy techniques are hot spot for the whole world. The conversion of air pollutants into high performance electrode for energy storage devices is a promising solution. In our previous work, we have used $H_2S$ as the reducing agent to reduce graphene oxide (GO) into graphene and obtained a S/rGO composite with great potential as cathode material for Li-S battery. Following this innovative strategy, we successfully optimized the interaction between $H_2S$ and GO, and a high-density S/graphene monolith was prepared with abundant "ink-bottle like" meso-pores□leading to a high density of the electrode and excellent confinement of polysulfides, thus a high volumetric capacity was delivered for compact Li-S battery. Besides the use of $H_2S$, we developed the traditional textbook reaction between $H_2S$ and $SO_2$ into a totally green and fast reaction in water medium to prepare highly-dispersible sulfur nanoparticles (HDS) with tuneable particle size and high stability. The HDS is contaminant-free, solution processable and size-controllable, and is thus ideal for achieving solution hybridization with a conducting matrix to prepare sulfur/carbon materials for different functions in Li-S battery. We have used Li-S batteries as a typical system to demonstrate the advantages of HDS as both an electrode material and an interlayer. Our approach shows the promise to convert air pollutants into useful materials for Li-S battery and other future applications.

## MATERIALS AND METHODS

### Material

*Preparation of HDS and HDS/carbon composite.* Highly-dispersible sulfur nanoparticles were synthesized by the reaction between $H_2S$ and $SO_2$ in water. The hybridization of HDS with other conductive components (CNT, graphene, PANI, etc.) was realized by a simple solution-mixing of different components following a centrifugation process.

## RESULTS AND DISCUSSION

The HDS prepared show appealing intrinsic electrochemical performance when mixed with commercial super P conductive, inspiring us to use HDS in high-performance Li-S batteries. We have fabricated HDS/MWCNT composite with a high sulfur content and a low content (5%) respectively, and these two membranes play different roles in Li-S battery. HDS/MWCNT membrane with a high sulfur content can serve as a high-performance cathode, which is current collector and binder-free. The theoretical capacity can be achieved for the first discharge, indicating the high activity and utilization of HDS. Besides the use in cathode system, HDS/MWCNT membrane with a low sulfur content is also used as the interlayer to reduce the shuttle effect, and the HDS in the interlayer have chemical affinity towards polysulfides, and thus resulting in an improved a cycle performance and rate performance. Moreover, the cathode we constructed also shows promising volumetric performance, which can pave the practical applications of our approach.

## CONCLUSIONS

$H_2S$ and related air pollutants are developed to prepare novel energy storage materials. Specifically, we have used Li-S batteries as a typical system to demonstrate the advantages of HDS as both an electrode material and an interlayer. Our approach shows the promise to convert major air pollutants, into useful materials for Li-S battery and other future applications. Moreover, it helps realize a perfect marriage of pollutant control and a new energy storage technology.

## REFERENCES

1. Zhang, C., Lv, W., Yang, Q. –H., et al (2014). "Reduction of Graphene Oxide by Hydrogen Sulfide: A Promising Strategy for Pollutant Control and as an Electrode for Li-S Batteries", *Adv. Energy. Mater.*, Vol.4, p.1301565
2. Zhang, C., Liu, D. H., Yang, Q. –H., et al (2015). "A high-density graphene–sulfur assembly: a promising cathode for compact Li–S batteries", *Nanoscale*, Vol.7, p.5592-5597
3. Liang, J. C., Zhang, C., Yang, Q. –H., et al (2017). "From Trash to Treasure: Turning Air Pollutants into Materials for Energy Storage", *ChemNanoMat*, DOI: 10.1002/cnma.201700038
4. Liu, D. H., Zhang, C., Yang, Q. –H., et al (2017). "Conversion of air pollutants into high performance S cathode in Li-S batteries", *In revision*

# Biomass-based Three-Dimensional Network of Carbon Nanosheets for Lithium-Ion Battery Anodes

Shasha Guo[1], Yaxin Chen[1], Xiaohong Chen[1], Jisheng Zhou[1], Zhaokun Ma[1], Huaihe Song[1]*

[1]State Key Laboratory of Chemical Resource Engineering, Beijing Key Laboratory of Electrochemical Process and Technology for Materials, Beijing University of Chemical Technology, Beijing 100029, P. R. China
*songhh@mail.buct.edu.cn

## INTRODUCTION

Carbon nanosheets have been the ideal electrode materials for electrochemical storage devices due to their high specific surface area, chemical stability, and electric conductivity[1]. What's more, it is well-known that the N-doping of carbon nanosheets can enhance the electrochemical properties of carbon materials by introducing more active site. Under the consideration of environmental issues and low-cost, researchers pay more attention to the carbon source of sustainable biomass[2]. In addition, many biomass materials are rich in heteroatoms themselves and can be act as the ideal carbon precursors. In this work, N-doped three-dimensional porous carbon ultrathin nanosheets (N-CNS) were fabricated by template method using soybean milk as the carbon source. When used as anode materials for the lithium ion batteries, the obtained carbons exhibit excellent electrochemical performance.

## MATERIALS AND METHODS

All the reagents were AR grade and commercial available without further purification. During the synthesis process, NaCl was added into soybean milk colloid to form a homogeneous solution. The solution was frozen-dried and the resulting white powder was annealed at 600 °C for 2 h under flowing nitrogen atmosphere. The obtained power was treated with deionized-water to remove the NaCl template and the final product was named as N-CNS.

## RESULTS AND DISCUSSION

As presented in Fig (a-d), the N-CNS were obtained by the carbonization of soybean milk. The N-CNS consist of a mass of monodispersed and overlapped nanosheets with a thickness of about 10nm and a small amount of mesopores was observed that may be derived from the tiny NaCl particles embedded in the organic compounds during the carbonization process. When the N-CNS were used as anode materials of lithium ion battery, it exhibits a typical cyclic voltammogram curve (CV) of carbon anode materials with a pronounced cathodic peak at 0-1V during the 1$^{st}$ cycle and at 0-0.3V during the 2$^{nd}$ and 3$^{rd}$ cycle. The intensity of this peak in 1$^{st}$ cycle is much stronger than followed cycles (Fig e). These differences are related to the formation of solid electrolyte interphase (SEI) layer and the decomposition of electrolyte. Fig (f-g) show the electrochemical performance of N-CNS within the voltage range of 0.01-3V. To be specific, the first reversible specific capacity is 1334.3 mAh $g^{-1}$ at the current density of 50 mA $g^{-1}$. And the anode delivers excellent specific capacities of 1212, 555 and 336 mAh $g^{-1}$ corresponding to current densities of 50, 500, and 2000 mA $g^{-1}$, respectively (Fig f), and a very high reversible capacity of 355 mAh $g^{-1}$ at a current density of 1 A $g^{-1}$ after 500 cycles (Fig g).

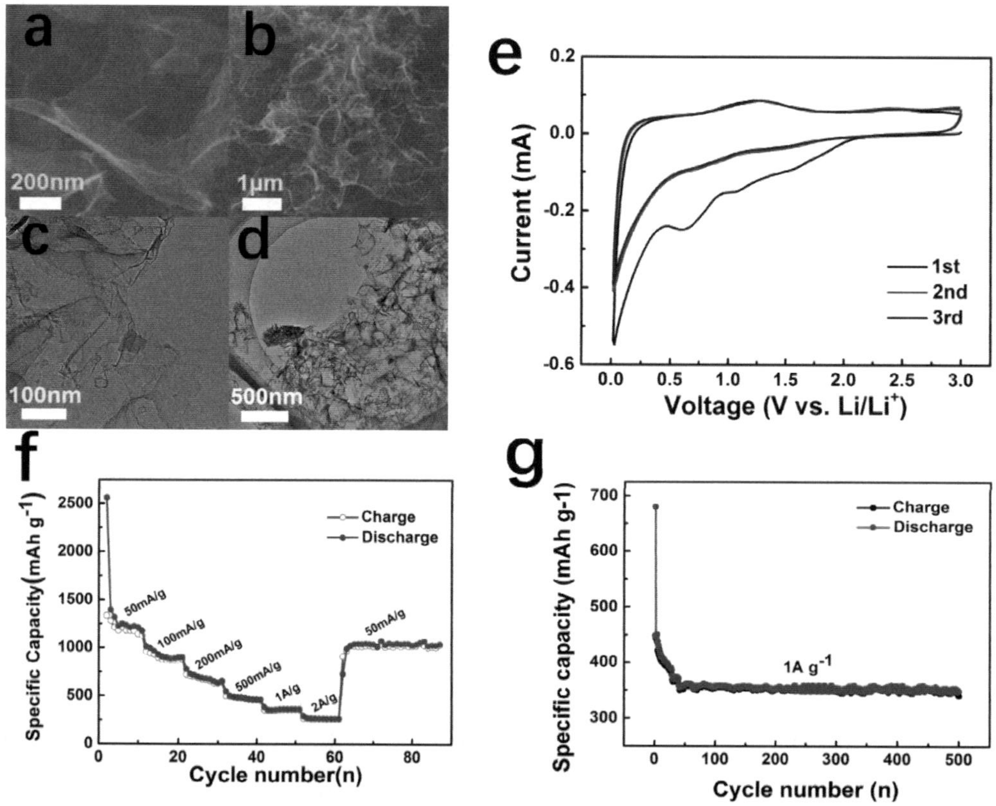


**Fig. (a-b) SEM images; (c-d) TEM images; (e) cyclic voltammetry curve of N-CNS at the scan rate of 0.1 mV $s^{-1}$; (f) rate performance of N-CNS at the current densities of 50, 100, 200, 500, 1000, and 2000 mA $g^{-1}$; (g) cyclic performance of N-CNS at the current densities of 1A $g^{-1}$.**

## CONCLUSIONS

The N-CNS is prepared by an easy, green, and scalable strategy with the assistance of NaCl. Our results show that biomass-based N-CNS are very favorable for the fast and steady transfer of electrons and ions and thus show good electrochemical performance.

## REFERENCES

1. Zheng, X., Luo, J. and Yang, Q. (2015). "Two-Dimensional Porous Carbon: Synthesis and Ion-Transport Properties", Adv Mater, Vol.27(36): p. 5388-5395

2. Roberts, A.D., Li, X. and Zhang, H. (2014). "Porous carbon spheres and monoliths: morphology control, pore size tuning and their applications as Li-ion battery anode materials", Chem Soc Rev, Vol.43(13): p. 4341-4356

# Carbon Matrix for C/C Composite by Film Boiling Technique

Yasushi Soneda[1*], Norio Iwashita[1], Michimasa Uda[2], Hiroshi Yamauchi[2]

[1] National Institute of Advanced Industrial Science and Technology (AIST), Tsukuba, Ibaraki, Japan
[2] IHI Aerospace, Tomioka, Gumma, Japan
*y.soneda@aist.go.jp

## INTRODUCTION

During the fabrication of carbon fiber reinforced carbon composites (C/C composites), the carbon matrix are produced by the carbonization of impregnated organic materials, mainly pitch. For the densification of C/C composites, the repetition of pitch impregnation and the carbonization under HIP (Hot Isostatic Pressing) condition several times are necessary. Since these processes require the long fabrication time and largely affect to the production cost, the reduction of processing period is eagerly desired.

The new technique of matrix impregnation so-called the film boiling (FB) process was reported [1, 2], in which the carbon fiber preform was heated in organic solvent, such as cyclohexane to maintain FB condition and the matrix carbon was directly deposited from solvent vapor decomposition.

We have reported the superior graphitizability of carbons from FB technique on the graphite plate compare to the matrix carbons synthesized from pitch by HIP condition [2-4]. In this study, the effects of precursor solvents and pressure on the synthesis of matrix carbons by FB method were pursued.

## MATERIALS AND METHODS

For the synthesis of C/C test specimens by FB method, carbon fiber felt (150 mm dia. and 15 mm thick) was fixed between graphite holders and placed in a reaction vessel containing excess amount

of cyclohexane or toluene and then heated by induction heating ca. 1000°C. The reaction

temperature was monitored by a thermocouple attached inside of holder plates and the solvent vapor was refluxed by condenser. The reaction experiments were performed under atmospheric pressure or 0.5 MPa. After the several hours of reaction, carbon felt was filled with deposited carbon matrix. The carbon deposits between carbon felt and graphite holder were collected and examined its graphitization behaviour with further heat treatment by means of X-ray diffraction, Raman scattering and SEM observation.

## RESULTS AND DISCUSSION

**Fig.1** shows 002 X-ray diffraction pattern of matrix carbon by FB method. The raw samples were obtained at 1000 °C and then these carbons were further heat treated up to 2800°C to examine the graphitization behaviour. At the 1000 °C, the carbon from cyclohexane shows narrower diffraction than the carbons from toluene. In the case of toluene, the weight of deposited carbon is 2.5 times high in 0.5 MPa of reaction compare to the atmospheric pressure, though the crystallinity of carbons from both pressures is almost the same. By the post heat treatment, all the sample carbons showed the improvement of crystallinity and become similar diffraction lines around 1500 °C. At 2800°C, the matrix carbons from toluene showed better graphitization than the one from cyclohexane, indicated by higher angle diffraction positions with narrower line shape. These crystallinity changes of matrix carbons from different synthesis conditions (solvent and pressure) with heat treatment are also supported by Raman scattering profiles. At the 2800 °C, the carbon matrix from toluene with 0.5 MPa of pressure only shows the asymmetric 2D peak in Raman scattering, which is indicative for the formation of three dimensional graphite structure.

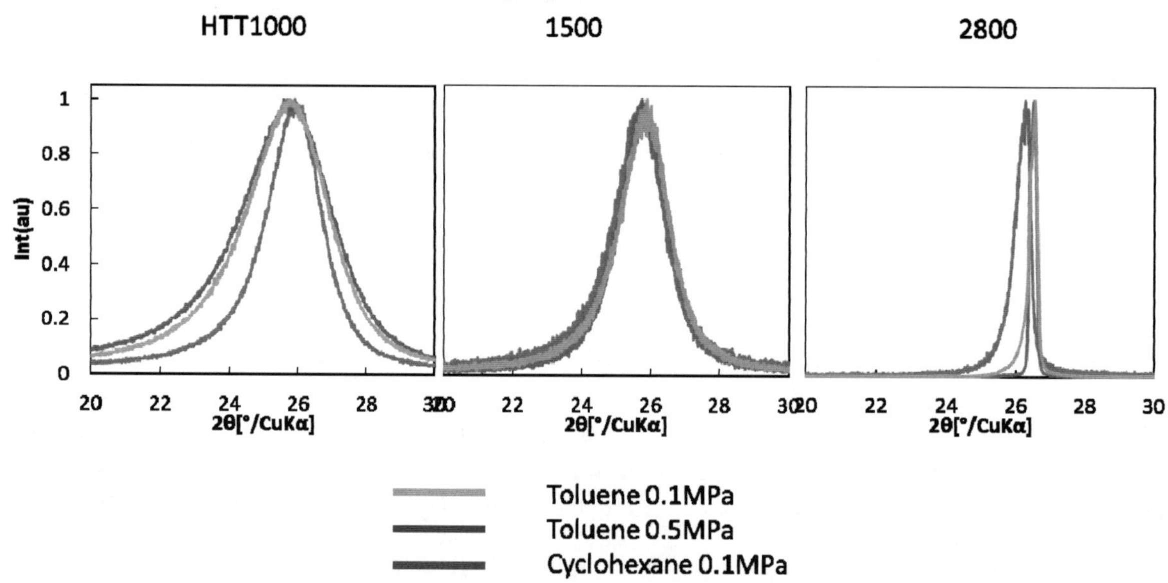


**Fig.1 002 X-ray diffraction pattern of matrix carbon by FB method.**

The crystal size along the *c*-axis of the FB carbons from toluene and cyclohexane was measured by 002 diffraction from X-ray powder pattern and plotted against the heat treatment temperature in **Fig.**

**2**. FB-carbon from toluene with 0.5Mpa shows marked increase in the crystal size than the carbons from toluene and cyclohexane with 0.1MPa.

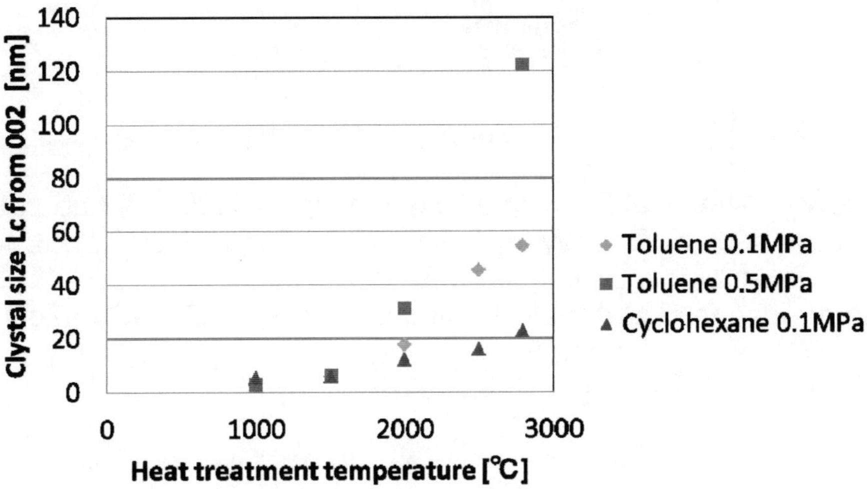


Fig.2 Crystal size Lc from 002 diffraction of matrix carbons by FB method.

## REFERENCES


1. Rovillain D., Trinquecoste M., Bruneton E., Derre A., David P., Delhaes P., Carbon 39 (2001)1355-65.
2. Soneda Y., Yamauchi H., Uda M., Abstract of 42th annual meeting of the Carbon Society of Japan (2015.12) 2C02, Osaka, Japan.
3. Yamauchi H. Uda M., Suzuki S., Soeda H., Soneda Y., Iwashita N., Abstract of JCCM17, (2016.3) Kyoto, Japan.
4. Soneda Y., Yamauchi H., Uda M., Extended Abstracts of Carbon2016, Penn State, (2016) P4-37.

# Carbon Nanofiber Film as Binder-Free Anode Materials for Sodium-Ion Storage

***Guo, X***[1], Zhou, J[1, 2]*, Song, H[1, 2], Chen, X[1]

[1] State Key Laboratory of Chemical Resource Engineering, Beijing Key Laboratory of Electrochemical Process and Technology for Materials, Beijing University of Chemical Technology, Beijing 100029, P. R. China
[2] Changzhou Institute of Advanced Materials, Beijing University of Chemical Technology, Jiangsu, P. R. China.
* zhoujs@mail.buct.edu.cn

## INTRODUCTION

Flexible energy storage devices with high energy and high power densities will be emerging power source to support flexible electronic devices with their development.[1] Just like flexible LIBs, flexible SIBs are also becoming a research hot topic for future energy storage. Recently, electrospinning has been developed to prepare fragmentary hard carbon nanofibers (CNFs) as anode materials for SIBs.[2] Moreover, Cross-linking is another effective strategy to enhance mechanical properties and electronic conductivity of electrospun nanofiber films so as to improve their application potential. [3] Herein we employed $Cu(NO_3)_2$ as an interesting cross-linking agent to prepared cross-linked carbon nanofibers using the cheaper precursor polyvinylpyrrolidone (PVP) by one-step electrospinning.

## MATERIALS AND METHODS

In a typical process, 0.62 g PVP and 0.17 g $Cu(NO_3)_2{\cdot}5H_2O$ was dissolved in mixed solution of 10ml ethanol and 1ml DMF to form a precursor solution by stirring at 60°C for 4 hours. During electrospinning, precursor solution was fed at a rate of 0.5 $mm{\cdot}min^{-1}$ at direct current voltage of 27 kV. Then fibers were dry in air at 150°C for 24 hours. Afterwards, the nanofibers were stabilized in Ar at 280°C for 2 hours, then the temperature was increased to 800°C and held for 2 hours with the propose of carbonizing PVP nanofibers. For comparison, the carbon nanofibers (CNFs) without cross-linked structure were obtained by annealing PVP fibers without $Cu(NO_3)_2{\cdot}5H_2O$ via a similar method.

## RESULTS AND DISCUSSION

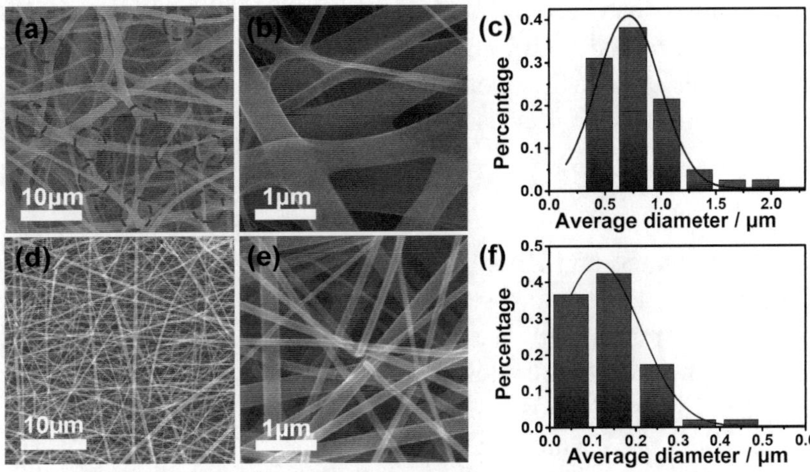


**Fig. 1 SEM image of CNFs (a and b) and CL- CNFs (c and d).**

Fig. 1a and b shows the SEM of CL-CNF film, and this film possesses lots of cross-linked structures as shown in red circles. And the cross-linked points are integrated by single carbon nanofiber. Moreover, the mainly diameter distribution of CL-CNFs are 0.7 μm (Fig. 1c). Compared with CL-CNF films, CNF films have smaller diameter in 0.1 μm (Fig. 1f). More importantly, as shown in Fig.1d and e, CNF films don't display cross-linked structure, which is very different from CL-CNF films.

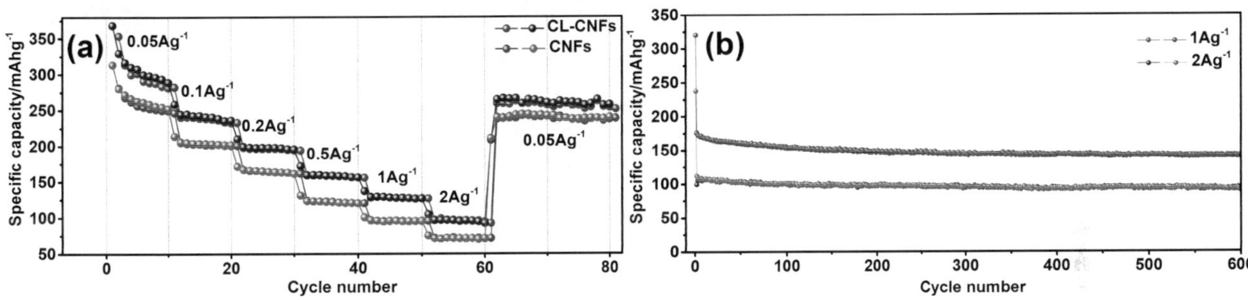


**Fig. 2 Rate performance of CL-CNFs films and CNFs films (a); long term cycling performace of CL-CNFs films at 1, 2 A $g^{-1}$ (b).**

Thereafter, the rate performances of CL-CNFs films and CNFs films are depicted in Fig. 2a. At a current of 50 mA $g^{-1}$, the CL-CNFs films delivered reversible capacity as high as 353.3 mA h $g^{-1}$. Then the current gradually increased to 100, 200, 500, 1000 and 2000 mA h $g^{-1}$ with 10 cycles at each current. And the corresponding reversible capacities can reach up to 240.4, 198.2, 159.2, 127.9 and 96.3 mA h $g^{-1}$, respectively. When current came back to 50 mA h $g^{-1}$, the reversible capacity still maintained 259.3 mA h $g^{-1}$, suggesting excellent stability of CL-CNFs films. What's more, the reversible capacities of CNFs films were inferior to CL-CNFs films, and the different capacity of the two films should be related to the cross-linked structure. To further estimate the long-term cycling ability of CL-CNFs films, the films were tested by galvanostatic discharged and charged at 1 A $g^{-1}$ and 2 A $g^{-1}$ (Fig. 2b). At 1 A $g^{-1}$ and 2 A $g^{-1}$, the CL-CNFs films maintain 141 mA h $g^{-1}$ and 94 mA h $g^{-1}$ after 500 cycles without obvious capacity decline. And after the loss stage, the coulombic efficiency approaches 100%, indicating the excellent capacity retention of CL-CNFs films.

## CONCLUSIONS

In summary, CNFs films with cross-linked structures were reasonably designed and successfully synthesized using a simple and convenient electrospinning method with PVP. Compared with CNFs films, CL-CNFs films were proved to be a promising anode, showing higher reversible capacity, superior rate capability and ultralong cycling stability. This can be ascribed to the cross-linked points. Therefore, this study will open an opportunity to create high performance carbon nanofibers as electrodes for sodium-ion batteries.

## REFERENCES


1. Li, L.?Wu Z., Yuan, S. and Zhang, X.-B. (2014). "Advances and challenges for flexible energy storage and conversion devices and systems", Energy Environ. Sci., Vol.7(7), p.2101

2. Li, W., Zeng, L., Yang, Z., Gu, L., Wang, J., Liu, X., Cheng, J. and Yu, Y. (2014). "Free-standing and binder-free sodium-ion electrodes with ultralong cycle life and high rate performance based on porous carbon nanofibers", Nanoscale, Vol.6(2), p.693-698

3. Cheng, Y.-L., Huang, L., Xiao, X., Yao, B., Yuan, L.-Y., Li, T.-Q., Hu, Z.-M., Wang, B., Wan, J. and Zhou, J. (2015). "Flexible and cross-linked N-doped carbon nanofiber network for high performance freestanding supercapacitor electrode", Nano Energy, Vol.15, p.66-74

# Carbon nanoparticle containing hydrogel nanocomposites with enhanced IR sensitivity

Krisztina László[1*], Barbara Berke[1,2], Lionel Porcar[2], Orsolya Czakkel[2],

[1]Department of Physical Chemistry and Materials Science, Budapest University of Technology and Economics, 1521, Budapest, Hungary
[2]Institute Laue Langevin, CS 20156, F – 38042 Grenoble Cedex 9, France.
*klaszlo@mail.bme.hu

## INTRODUCTION

Hydrogels with environmental sensitivity can provide a way to change some of their properties (i.e. size, optical properties, etc.) in a controlled way making a perfect candidate for actuators, sensors or vehicles for drug delivery systems. Their physical properties often restrict the applicability of pure polymer systems, which can be overcome by nanocomposite hydrogels.
Regardless of the targeted application, the comprehensive understanding and tuneability of the nature and kinetics of the response is inevitable. Our aim was to investigate the effect of quality and quantity of carbon nanoparticles with different structure and surface chemistry on the thermal response of nanocomposite system. Carbon nanotubes (CNTs) and graphene oxide (GO), respectively, were incorporated into the poly(N-isopropyl acrylamide) (PNIPAM) gel matrix. Because of their unique structure, outstanding mechanical and conductive properties, they can act not only as reinforcing agents, but may provide new type of sensitivity to the systems.

## MATERIALS AND METHODS

### Material

GO was obtained by the improved Hummers' method[1)]. CNT with improved water dispersibility were obtained by oxidizing commercial multi-wall CNT with cc. $HNO_3$[2)]. Pure and composite gels were synthesized via free radical polymerization at 20 °C[3)].

### Methods

The equilibrium swelling degree and elastic modulus of the samples were compared to scanning electron microscopy images. The temperature sensitivity of the samples were characterised by the kinetics of the temperature jump induced macroscopic shrinkage were studied on gel disks of 14 mm diameter with 20 → 50 °C temperature jump and their response to IR irradiation.

## RESULTS AND DISCUSSION

The observed effect of the two nanoparticles is significantly different. The mechanical properties of PNIPA gel were improved by incorporating GO, while the swelling degree of the GO@PNIPA systems significantly decreased. In contrast, CNT@PNIPA composites exhibited swelling and mechanical properties similar to those of pure PNIPA. the presence of CNPs appreciably modifies the kinetic response of the PNIPA gel. During the temperature-jump the behaviour of the composite gels is strongly affected by both the quality and quantity of the CNPs. CNT slightly accelerates the response of the pure PNIPA gel, and the resulted time constants correlate with the CNT content, however, a timeframe of even an order of magnitude longer was insufficient for the GO@PNIPA gels to reach the final equilibrium. Unlike CNT, the lowest GO content has the strongest effect on deswelling kinetics. Regarding the IR sensitivity tests, Fast shrinkage on exposure to IR laser irradiation, and quick recovery of the gels after exposure is observed in all cases (**Fig.1a**). A monotonic correlation can be found between the nanoparticle content and the temperature of the samples upon IR irradiation, but the concentration dependence was different for the two type of composite family (**Fig.1b**). The stronger IR absorption at low GO concentration seems to reveal that these complex processes suppress the IR responsivity of the hybrid.

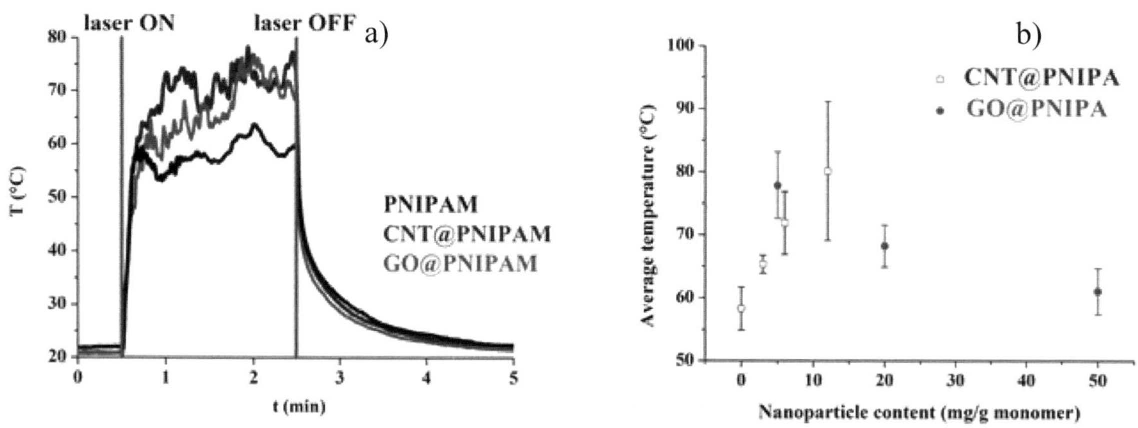


**Fig. 1. a)** Temperature (T) profile of PNIPA composites upon IR laser exposure (t: observation time) at 20 °C. **b)** Average temperature values of PNIPA composites during IR laser exposure

## CONCLUSIONS

Incorporation of CNT and GO results in different effects in the behaviour of the hybrids.
Their effect on the swelling degree and the elastic modulus shows a different trend. Due to the observed differences in the thermal responses of the systems, the time constant and the swelling ratio of the temperature-induced shrinkage can thus be adjusted by selecting the type and amount of nanoparticle loading. Both CNT and GO are able to enhance the infrared sensitivity of the PNIPA gel, thus opening a route for constructing novel drug transport and actuator systems. The difference of the influence of CNT and GO in the behaviours investigated may stem rather from their different chemical reactivity during the gel synthesis than from their different geometry.

## ACKNOWLEDGEMENT

The authors thank Prof. K. Kaneko for providing the graphite sample and Prof. T. Maekawa for his valuable advice in the IR experiments. The financial support of Hungarian grants OTKA K109558 is acknowledged. The work is part of the EU project FP7-PEOPLE-2010-IRSES-269267 and the Nanoporous and Nanostructured Materials for Medical Applications (Nanomed: H2020-MSCA-RISE-2016, project:734641).

## REFERENCES

1. Marcano, D., Kosynkin, D., Berlin, J. M., Sinitskii, A., Sun, Z. Z., Slesarev, A., Alemany, L. B., Lu W. and Tour, J. M. (2010). "Improved Synthesis of Graphene Oxide", *ACS Nano*, Vol.4, p.4806–4814

2. Wang, Z., Shirley, M. D., Meikle, S. T., Whitby, R. L. D. and Mikhalovsky, S. V. (2009). "The surface acidity of acid oxidised multi-walled carbon nanotubes and the influence of in-situ generated fulvic acids on their stability in aqueous dispersions" *Carbon,* Vol.47, p.73–79.
3. Manek, E., Berke, B., Miklósi, N., Sajbán, M., Domán, A., Fukuda, T., Czakkel, O. and László, K. (2016). "Thermal sensitivity of carbon nanotube and graphene oxide containing responsive hydrogels" *eXPRESS Polym. Lett.*, Vol.10, p.710–720.

# Carbon spheres with onion-like structure as an anode material for Lithium-ion batteries

Bin Cao, Huaihe Song, *Xiaohong Chen, Jisheng Zhou, Zhaokun Ma

State Key Laboratory of Chemical Resource Engineering, Beijing Key Laboratory of Electrochemical Process and Technology for Materials, Beijing University of Chemical Technology, Beijing 100029, China
Corresponding author: songhh@mail.buct.edu.cn

## INTRODUCTION

Commercial graphite-based anode materials have merits of low voltage hysteresis, high first coulombic efficiency and low lithium storage potential, which are suitable to match a cathode material for practical application. However, the capacity decrease is inevitable during cycling.[1] In order to overcome this shortage, various approaches, such as spheroidizing material particles and surface coating,[2] have been proposed, but the improvement still is limited because the post-treatment is temporary palliative and can not solve the problem essentially.

In brief, the high anisotropy of graphit is the primary reason for the structure instability. Accordingly, this point inspired us to find a suitable structure for carbon materials with less anisotropy. As an extension of the fullerene's family, carbon materials with an onion-like structure have stable performance as electrode materials due to their unique characteristics, such as high chemical and structural stability, and excellent electrical conductivity. Most importantly, carbon materials with onion-like structure have approximate isotropy due to the carbon layers arranged concentrically.

Herein, we report a novel graphitized carbon spheres (GCS) with an onion-like structure. The GCS has a reversible capacity of 310 mAh $g^{-1}$ that is the highest recording among the reported carbon materials with onion like structure and graphite-like electrochemical behavior.[3] The GCS also exhibits excellent long-term durability with 195 mAh $g^{-1}$ over 1000 cycles at 200 mA $g^{-1}$. Therefore, our results provide a new strategy for carbon anode materials with long-term cycle stability by using multi-shelled and onion-like hybrid structure.

## MATERIALS AND METHODS

The carbon spheres precursor were synthesized by a modified chemical vapor deposition at 900 °C using pyridine as precursor and the detailed preparation method has been reported by our previous work.[4] Then, the carbon spheres were further heat treated at 2800 °C under $N_2$ atmosphere for one hour to get GCS.

## RESULTS AND DISCUSSION

The outer surface of GCS (**Fig. 1a** and **b**) is constituted of carbon basal planes. The multi-shelled structure of GCS can be confirmed by TEM characterization (**Fig. 1c**). Intersections of polygonal surfaces can be well distinguished in **Fig. 1d**. The inset of **Fig. 1d** shows the selected area electron diffraction (SAED) pattern with clear diffraction ring, indicating high graphitization degree of GCS. Lithium insertion/extraction can be clearly observed by a sharp pair of redox peaks at 0.01-0.2 V in CV curves of GCS (**Fig. 1e**). The charge-discharge profiles for 1st, 10th, 50th and 100th cycles are measured at current density of 50 mA $g^{-1}$. GCS present a typical graphite-like charge-discharge behaviour (**Fig. 1f**) with a long and stable plateau below 0.25 V vs. Li+/Li, which is beneficial for outputting high working voltage when coupled with a cathode in Li-ion full cell.

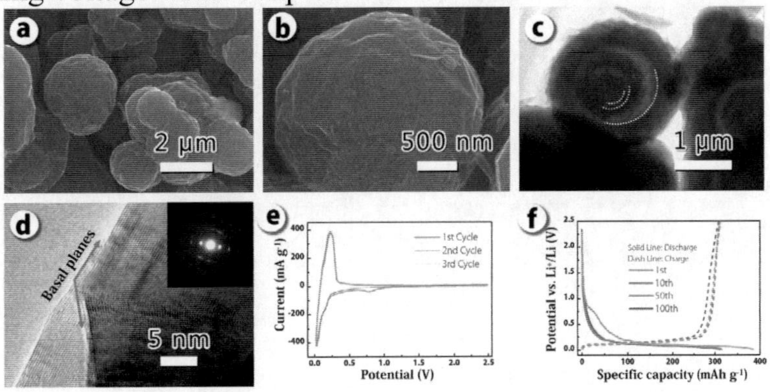


Figure 1. SEM images (a and b), HR-TEM images (c and d; inset: the selected area electron diffraction pattern), CV curves at 0.1 mV $s^{-1}$ (e) and Charge-discharge profiles at 50 mA $g^{-1}$ (f) for GCS.

After 100 cycles at 50 mA $g^{-1}$, GCS have a capacity retention of 99% almost without any capacity fade (**Fig. 2a**.). The semicircles decreased gradually after cycling and almost kept stable after 10th cycle, suggesting a few cycles maybe necessary to form a stable electrode/electrolyte interface (**Fig. 2b**). **Fig. 2c** shows the rate performance of GCS in the current density of 30 mA $g^{-1}$ to 2A $g^{-1}$. Moreover, the electrodes of GCS were cycled at 200 mA $g^{-1}$ for 1000 cycles. GCS have higher capacity retention of 91% with reversible capacity of 195 mAh $g^{-1}$.

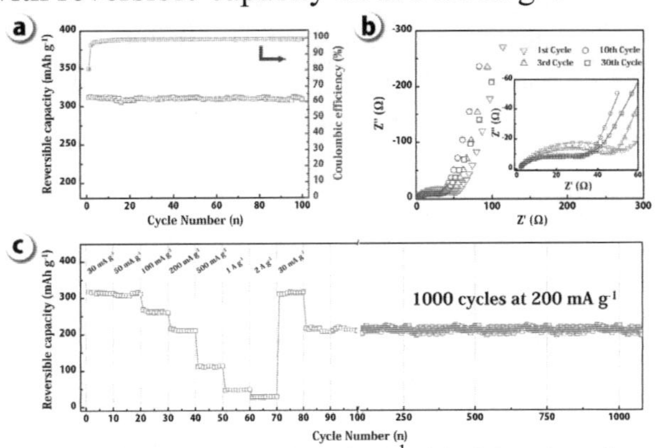


Figure 3. Cyclability and coulombic efficiency at 50 mA $g^{-1}$ (a), Nyquist plots for different cycles under full delithiation state (b) and rate performance (c).

## CONCLUSIONS

A novel graphitized carbon spheres (GCS) with an onion-like structure was prepared by a normal high temperature heat treatment without any post treatment. Since both multi-shelled and onion-like structure favor for improving cycle performance, GCS has an unexpected superior capacity retention of 91% after 1000 cycles at 200 mA $g^{-1}$.

## ACKNOWLEDGMENTS

This work was supported by the National Natural Science Foundation of China (U1610252), and Foundation of Excellent Doctoral Dissertation of Beijing City (YB20121001001).

## REFERENCES

1. Flandrois, S., Simon, B. Carbon materials for lithium-ion rechargeable batteries. *Carbon*, 1999, 37(2): 165-180.
2. Li, H., Zhou, H. Enhancing the performances of Li-ion batteries by carbon-coating: present and future. *Chem. Commun.*, 2012, 48(9): 1201.
3. Wang, H., Abe, T., Maruyama, S. et al. Graphitized Carbon Nanobeads with an Onion Texture as a Lithium-Ion Battery Negative Electrode for High-Rate Use. *Adv. Mater.*, 2005, 17(23): 2857-2860.

4. Cao, B.,Liu, H.,Xing, Z. et al. Preparation of Nitrogen-Doped Carbon Spheres by Injecting Pyrolysis of Pyridine. *ACS Sustain. Chem. Eng.*, 2015, 3(8): 1786-1793.

# Carbon Support Effects on Catalysts for Hydrogenation of Multifunctional Chemicals

Radhika Rao[1], Raoul Blume[2], Thomas Hansen[3], Erika Fuentes[4], Kathleen Dreyer[5], David Hibbitts[5], Yves Chabal[4], Robert Schlögl[2], Jean-Philippe Tessonnier[1*]

[1]Department of Chemical and Biological Engineering, Iowa State University, Ames, USA
[2]Max Planck Institute for Chemical Energy Conversion, Berlin, Germany
[3]Center for Electron Nanoscopy, Technical University of Denmark, Lyngby, Denmark
[4]Materials Science and Engineering Physics, University of Texas Dallas, Richardson, USA
[5]Department of Chemical Engineering, University of Florida, Gainesville, USA
*tesso@iastate.edu

## INTRODUCTION

Carbon-supported noble metal catalysts are widely used for hydrogenation reactions to produce specialty and fine chemicals. Although carbons are typically assumed to be inert supports, their physico-chemical properties are known to influence the catalytic performance of noble metal nanoparticles decorating their surface. Currently, these phenomena are attributed to metal-support interaction (MSI) effects. While the exact nature of these interactions remains elusive in the catalysis literature, the optoelectronic field provides deeper insights into the properties of carbons. Studies from photonics demonstrate that the optical and electronic properties of carbons vary significantly with graphitization and functionalization[1)]. This understanding suggests that carbon scaffolds might be candidates for electronic metal support interaction (EMSI) effects.

Previous attempts in understanding MSI effects have failed due to the heterogeneous nanoscale structure of the carbon scaffolds[2)]. Our work is built on the hypothesis that structure-activity correlations for carbon-supported catalysts can be studied systematically if the support structure is controlled at the nanoscale. This study attempts to identify the role of graphitization, functional group compositions, and atom hybridization on the MSI effects by decreasing the support complexity and de-convolving the chemical and electronic effects.

## MATERIALS AND METHODS

In order to systematically study the effect of carbon materials on the supported metal active phase, stacked cup carbon nanotubes (SCCNTs) were used as supports in this work. The dual structure of SCCNTs consisting of a graphitic core and an amorphous carbon shell allowed us to obtain two support series with controlled graphitization and functionalization using simple thermochemical treatments. Pd nanoparticles were deposited on their surface using incipient wetness impregnation and the catalysts were tested for the liquid phase hydrogenation of α,β-unsaturated aldehyde. Detailed characterization was conducted using Raman spectroscopy, X-ray diffraction (XRD), hydrogen chemisorption adsorption isotherms, high pressure Fourier transform infrared spectroscopy (FT-IR), synchrotron based X-ray photoelectron spectroscopy (XPS), and aberration corrected transmission electron microscopy (TEM). The experimental results were supported with theoretical calculations using density functional theory (DFT) studies.

## RESULTS AND DISCUSSION

Clear trends in hydrocinnamaldehyde (HCALD) selectivity were observed with the support treatment temperature up to 800°C (Fig. 1). The selectivities were compared at iso-conversion in a kinetically controlled regime and the turnover frequencies for all catalysts were similar. XRD, hydrogen chemisorption, and TEM confirmed the presence of large (~5nm) Pd particles on all supports. The bulk of the nanoparticles consisted of Pd in the metallic state ($Pd^0$) as demonstrated by XPS and aberration corrected TEM. Synchrotron based XPS revealed the presence of an additional $Pd^{\delta+}$ phase in all the catalysts. The ratio of $Pd^0/Pd^{\delta+}$ was observed to be in coherence with HCALD selectivity along with the support properties, demonstrating the presence of interfacial charge transfer between the metal and supports. Evidence for EMSI was strengthened by the differences in adsorption properties of the Pd metal active phase as confirmed by hydrogen chemisorption isotherms. The experimental results were further supported by DFT studies that provided deeper insights into EMSI in Pd/C catalysts and their correlations with support properties.

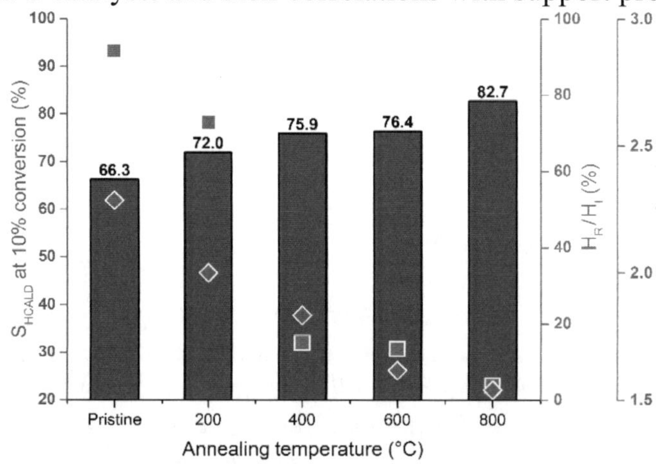


**Fig. 1.** HCALD selectivity ■ at 10% conversion for Pd/C catalysts supported on carbon materials annealed at different temperatures; Ratio of reversibly and irreversibly bound hydrogen from hydrogen chemisorption ($H_R/H_I$) ▪; Ratio of peak intensity for Pd in metallic state to δ+ charged state as observed with synchrotron based XPS ($Pd^0/Pd^{\delta+}$) ◆.

## CONCLUSIONS

The correlations between support properties and EMSI effects established in this work demonstrate that contrary to the common assumption, carbon materials are not inert supports. The observed support effects are significant and they lead to interfacial charge transfer represented by the $Pd^{\delta+}$ phase even for large (~5 nm) particles which would typically be expected to demonstrate bulk-like behaviour. Our work demonstrates how simple thermochemical treatments can be used to control EMSI and thereby opens new avenues for the rational design of Pd/C catalysts.

## REFERENCES


1. Loh, K. P., Bao, Q., Eda, G. and Chhowalla, M. (2010). "Graphene oxide as a chemically tunable platform for optical applications", *Nat. Chem.* Vol.2, p.1015-1024

2. Su, D. S., Perathoner, S. and Centi, G. (2013). "Nanocarbons for the development of advanced catalysts" *Chem. Rev.* Vol.113, p.5782-5816

# $Co_{0.85}Se$ Nanosheets/Graphene Composite Film as Binder-free Anode for Lithium-ion Batteries

Zhang, G[1], Zhou, J[1, 2]*, Chen X[1], Song, H[1, 2]

[1]State Key Laboratory of Chemical Resource Engineering, Beijing Key Laboratory of Electrochemical Process and Technology for Materials, Beijing University of Chemical Technology, Beijing 100029, P. R. China
[2]Changzhou Institute of Advanced Materials, Beijing University of Chemical Technology, Jiangsu, P. R. China
*zhoujs@mail.buct.edu.cn

## INTRODUCTION

Nowadays, developing flexible energy storage devices such as flexible lithium ion batteries (LIBs) with high performance are urgently needed in the next generation, flexible and wearable electronic devices[1, 2]. Transitional metal selenides (TMSs) have attracted interest as anode materials for lithium-ion batteries (LIBs) due to their high capacities, thermal and chemical stability. Cobalt selenide, as one of most important TMSs, has been investigated as anode materials for LIBs. However, cobalt selenide anodes possess some drawbacks such as large volume expansion during the lithiation/delithiation process and poor cycle stability. Graphene has attracted significant attention in flexible electrodes field due to its unique structure, high electrical conductivity and excellent mechanical properties[3, 4]. Here, flexible free-standing $Co_{0.85}Se$ nanosheets (NSs)/graphene composite film was prepared as anode of LIBs with excellent electrochemical performances.

## MATERIALS AND METHODS

In a typical procedure, $Co_{0.85}Se$ NSs were prepared by a solvothermal method. Then 10 mg $Co_{0.85}Se$ NSs were dispersed in DIW by sonication for 10 minutes, then a desired amount of the GO dispersion was then added to the suspension of $Co_{0.85}Se$ NSs. The mixture of $Co_{0.85}Se$ NSs and GO was then sonicated for 60 minutes and vacuum-filtered. After filtering, annealing at 600°C for 2 h in an argon gas atmosphere.

## RESULTS AND DISCUSSION

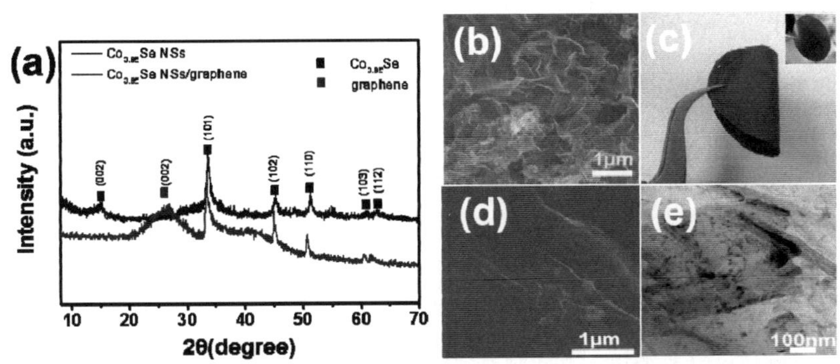


**Fig. 1 (a) XRD patterns of $Co_{0.85}Se$ and $Co_{0.85}Se$NSs/graphene film, (b) SEM image of $Co_{0.85}Se$ NSs, optical images of $Co_{0.85}Se$ NSs/graphene film and $Co_{0.85}Se$ NSs/GO film (inset one), SEM image (d) and TEM image of $Co_{0.85}Se$ NSs/graphene film.**

In the XRD pattern of $Co_{0.85}Se$ (Fig. 1a), diffraction peaks match very well with standard $Co_{0.85}Se$ (JCPDS No.52-1008). The weak diffraction peak appeared at 15.4° should be attributed to the (002) crystal plane, implying the presence of layered structural $Co_{0.85}Se$. The TEM images (Fig. 1b) presents that the $Co_{0.85}Se$ owns wave-like two dimensional (2D) nanosheets structure. By simple mixing, filtration and annealing, $Co_{0.85}Se$ NSs/graphene film was prepared. In the XRD pattern of $Co_{0.85}Se$ NSs/graphene film (Fig. 1a), a broadened diffraction peak appears at 25°, which can be indexed into the disorderedly stacked graphene, indicating the successful reduction of GO, and all of the other diffraction peaks can still be attributed to $Co_{0.85}Se$, indicating high thermal stability of $Co_{0.85}Se$. In Fig. 1c the optical image indicates that $Co_{0.85}Se$/graphene film has good flexibility. The SEM and TEM images (Fig. 1d and e) show that $Co_{0.85}Se$ NSs are uniformly incorporated between the graphene layers, resulting in sandwiched structures.

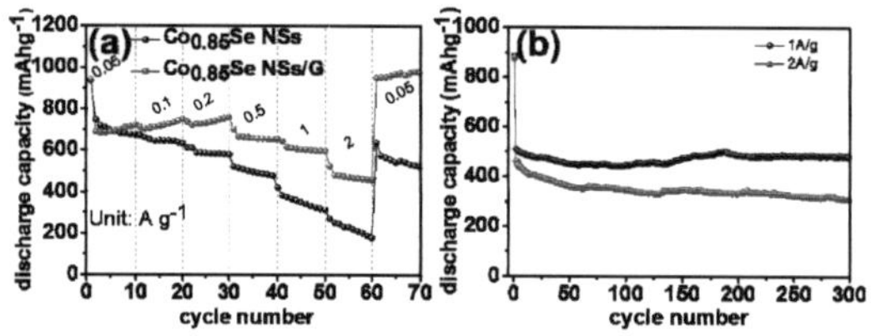


**Fig. 2 (a) rate performance of $Co_{0.85}Se$ NSs and $Co_{0.85}Se$ NSs/graphene, (b) cyclic properties of $Co_{0.85}Se$ NSs/graphene at high current density of 1 and 2A/g.**

Subsequently, the electrochemical performances of $Co_{0.85}Se$ NSs/graphene film and the bare $Co_{0.85}Se$ NSs were evaluated as anode materials for LIBs. The $Co_{0.85}Se$ NSs/graphene film exhibits better rate performance than $Co_{0.85}Se$ NSs in Fig. 2 a. From the Fig. 2b, we can see the $Co_{0.85}Se$/graphene composite has better cycling performances at high current, the discharge capacity can keep at 480 and 310 mA h $g^{-1}$ after 300 cycles at 1 and 2 A $g^{-1}$, respectively. The excellent electrochemical performances of $Co_{0.85}Se$ NSs/graphene film are attributed to the elastic graphene matrix, which not only provides a high electrical conductivity pathway, but also accommodates the volume changes and prevents the breakdown electrode.[5]

## CONCLUSIONS

In summary, a flexible $Co_{0.85}Se$ NSs/graphene film has been prepared by a vacuum filtration method and thermal reduction process. And this film as binder-free anode for LIBs has excellent electrochemical performances. The highly conductive graphene not only provides enhanced electrical conductivity of the whole electrode, but also acts as an elastic layer to buffer the volumetric expansion of $Co_{0.85}Se$ nanosheets. The unique structure of graphene and $Co_{0.85}Se$ NSs composite film resulted in fast Li ion diffusion and excellent rate performance and cyclic stability than pure $Co_{0.85}Se$. Accordingly, this design provides a promising candidate for binder-free flexible anode for energy storage.

## REFERENCES

1. Bruce, P. G., Scrosati, B. and Tarascon, J. M. (2008). "Nanomaterials for Rechargeable Lithium Batteries", *Angew. Chem. Int. Ed.*, Vol.47, p.2930-2946

2. Gwon, H., Kim, H.-S., Lee, K. U., Seo, D.-H., Park, Y. C., Lee, Y.-S., Ahn, B. T. and Kang, K. (2011). "Flexible energy storage devices based on graphene paper", *Energy Environ. Sci.,* Vol.4, p.1277-1283

3. Wang, R., Xu, C., Sun, J., Gao, L. and Lin, C. (2013). "Flexible free-standing hollow $Fe_3O_4$/graphene hybrid films for lithium-ion batteries", *J. Mater. Chem. A,* Vol.1, p.1794-1800

4. Prabakar, S. J., Hwang, Y.-H., Bae, E.-G., Shim, S., Kim, D., Lah, M. S., Sohn, K.-S. and Pyo, M. (2013). "$SnO_2$/Graphene Composites with Self-Assembled Alternating Oxide and Amine Layers for High Li-Storage and Excellent Stability", *Adv. Mater.*, Vol.25, p.3307-3312

5. Wang, D., Kou, R., Choi, D., Yang, Z., Nie, Z., Li, J., Saraf, L. V., H, D., Z, J., Graff, G.L., Liu, J., Pope, M. A. and Aksay, I. A. (2015). "Ternary Self-Assembly of Ordered Metal Oxide Graphene Nanocomposites for Electrochemical Energy Storage", *ACS Nano*, Vol.4, p.1587-1595

# CONFIRMATION OF THE FORMATION OF OXIDATIVE DEBRIS DURING THE HUMMERS-OFFEMAN REACTION IN THE SYNTHESIS OF GRAPHENE OXIDE

G. Ramos-Fernández[1*], J. Pérez-Galera, D. Domene López[1], I.Martin-Gullón[1]
[1]Department of Chemical Engineering, University of Alicante, Spain
*gloria.ramos@ua.es

## INTRODUCTION

The structure of graphene oxide (GO) is a matter of discussion. A two-component structural model for GO was proposed as a direct product of the oxidation reaction: slightly oxidized, structurally intact graphene-like sheets (bwGO) and highly oxidized molecular species, oxidative Debris (OD), adsorbed on it, which can be separated by a base digestion [1]. OD is a polyaromatic oxidized material, produced during the oxidative transformation of graphite into GO. However, the origin of OD is still in dispute. Moreover, characterizing the structure and location of supposed OD on nanosheets of GO produced during the oxidations process is difficult.

Recently, this model was strongly criticized stating that the OD is formed when producing the base digestion of previously dried graphite oxide and not in the Hummers-Offemann reaction [2]. The presence of OD complicates the structures and properties of GO, it is important to confirm the origin and structure of OD. Futhermore, identifying the locations and the origin of OD on graphene oxide nanosheets has yet to be explored, which is of particular significance for obtaining deep insights regarding the surface chemistry of graphene oxide and for further functionalization and applications of graphene oxide.

In this study, determines the amount of bwGO and OD (including humic and fulvic like fractions), and its spectroscopic and electron microscopy characterization, formed when starting from the Hummers-Offerman reaction, and just after the base digestions was carried out without any intermediate drying of graphite oxide nor sonication treatment.

## MATERIALS AND METHODS

Two different samples of graphene oxide were studied. The first one is a conventional graphene oxide prepared using a modified Hummers-Offeman method [3], denoting this as GO. The second sample is based on the previous one but included at the end a simple ammonia wash at room temperature (and further filtration) to remove part of the OD, denoted as aGO. Both GO and aGO were digested with concentrated aqueous NaOH and NH3 up to 1 hour at 70ºC with subsequent centrifugation [4]. The pale-yellow supernantants of these centrifugation steps containing the OD, including fulvic and humic substances, and the sediment was the bwGO.

## RESULTS AND DISCUSSION

The Fig. 1 shows differences between GO sheet, bwGO, humic acid and OD fraction obtained after separation. Fig. 1 A, show a representative GO sheet with a flat geometry. Additionally, many GO sheets presented a surface fully covered by darker spots, of quite small size. According to Chen and Chen [5], these spots which also cover the TEM carbon lacey grid, correspondd to oxidated debris anchored by py-py stacking on aromatic islands of bigger GO sheets. On the other hand, the aGO sample (Fig 1.b) preserves laminar graphene-like structures, but more aggregated and wrapped than the original GO, as can be expected according to loss of OD. In some cases, tiny spots were also detected, although in a much lower extension. Fig. 1C shows bwGO sheet, preserves laminar graphene-like structures, but more aggregated and wrapped than the original GO. On the contrary, the humics (Fig 1.D) and OD-fraction (**Error! Reference source not found.**E) is an amorphous material with non-laminar structure, and it appears like an agglomeration of macromolecules.

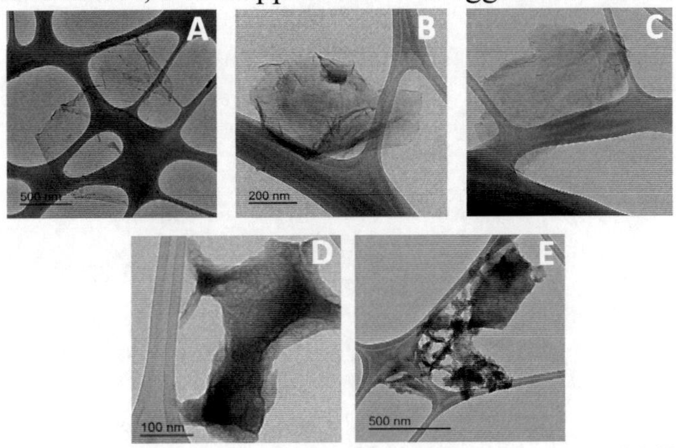


**Fig 1. TEM images of GO (A), aGO (B), bwGO (C), Humic acid (D) and OD (E).**

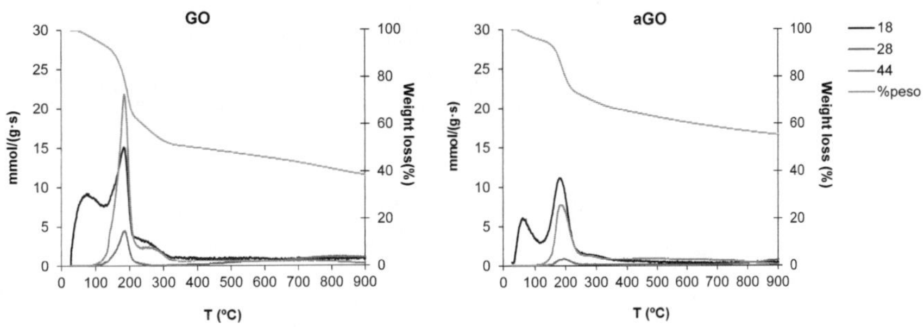


**Fig.2. TG-MS of GO and aGO.**

Figure 2, show the TG-MS plot and the corresponding quantitative analysis of evolven CO, $CO_2$ and $H_2O$ analysis; they reflect the differences in the surface chemistry between the GO and aGO. The most prominent weight loss is observed at around 180 ºC with the evolution of CO, $CO_2$ and $H_2O$, followed by another loss at 225ºC, and finally an extended tail. However, the TG plot is different for aGO, with a single and less pronounced weight loss at 180 ºC. This fact leads us to the conclusion that gases evolved from aGO correspond mostly to the decomposition of surperficial oxygen groups anchored to sheets, while in GO, they correspond to both anchored oxygen groups and to fragments of other hydrocarbon molecules and $SO_2$, liberated from the evolution of oxidative debris attached to the GO sheets. These results suggest that the base washing step has been effective and the oxidative debris has been easily removed in a big extent.

## CONCLUSIONS

We concluded that the OD formation is directly related to the pattern material and the oxidation process employed and does not depend on the subsequent processes such as the ultrasounds or the basic treatments used to separate the OD fraction.

## REFERENCES

[1] Rourke JP, Pandey PA, Moore JJ, Bates M, Kinloch IA, Young RJ, et al. The real graphene oxide revealed: Stripping the oxidative debris from the graphene-like sheets. Angewandte Chemie - International Edition. 2011;50(14):3173-3177.

[2] Dimiev AM, Polson TA. Contesting the two-component structural model of graphene oxide and reexamining the chemistry of graphene oxide in basic media. Carbon. 2015;93:544-554.

[3] Rodriguez-Pastor I, Ramos-Fernandez G, Varela-Rizo H, Terrones M, Martin-Gullon I. Towards the understanding of the graphene oxide structure: How to control the formation of humic- and fulvic-like oxidized debris. Carbon. 2015;84(1):299-309.

[4] Thomas HR, Day SP, Woodruff WE, Vallés C, Young RJ, Kinloch IA, et al. Deoxygenation of graphene oxide: Reduction or cleaning? Chemistry of Materials. 2013;25(18):3580-3588.

[5] Chen X, Chen B. Direct Observation, Molecular Structure, and Location of Oxidation Debris on Graphene Oxide Nanosheets. Environmental Science and Technology. 2016;50(16):8568-8577.

# Construction of integrated $RGO/MnFe_2O_4/g\text{-}C_3N_4$ catalyst for decomposition of methylene blue

Jiangying Qu[1], Xiyue Peng[1], Chuang Geng[1], Feng Gao*[1], and Jieshan Qiu[2]*

[1] Faculty of Chemistry and Chemical Engineering, Liaoning Normal University, Dalian, 116029, China.

[2] Carbon Research Laboratory, Center for Nano Materials and Science, School of Chemical Engineering, State Key Lab of Fine Chemicals, Dalian University of Technology, Dalian, 116024, China

## INTRODUCTION

Heterogeneous Fenton-like technology is an advanced oxidation method to deal with the organic dyes in aqueous environment. Double transition metal oxides with a general formula of $AB_2O_4$ (A, B=Mn, Fe, etc.) have made much contribution in Fenton-like process. Typically, $MnFe_2O_4$ shows great potential for the high catalytic and magnetic performance [1]. However, bare $MnFe_2O_4$ particles often suffer from aggregation. Graphene oxide (GO) selected as the precursor to prepare reduced graphene oxide (RGO) based RGO/metal oxide nanocomposites, wherein the residual oxygen-containing functional groups of RGO can firmly combine with the metal cations by strong interfacial interactions [2]. Furthermore, the graphitic carbon nitride ($g\text{-}C_3N_4$) opens up new opportunities for fabrication of $RGO/MnFe_2O_4/g\text{-}C_3N_4$ hybrid catalysts with enhanced photocatalytic efficiency [3]. Herein, we have proposed an effective strategy with hydrothermal method to fabricate stable $RGO/MnFe_2O_4/g\text{-}C_3N_4$ catalyst for the decomposition of MB dye. The synergetic effects of $RGO/MnFe_2O_4/g\text{-}C_3N_4$ hybrid have brought the high catalysis performance on MB decomposition.

## MATERIALS AND METHODS

*Synthesis of $RGO/MnFe_2O_4/g\text{-}C_3N_4$ hybrids*

Homogeneous $GO/MnSO_4$ suspension derived from the modified Hummers method. $FeSO_4{\cdot}7H_2O$ and $g\text{-}C_3N_4$ were added into the above suspension, 2 mol $L^{-1}$ KOH was then added until pH=12. The system was then transferred into a Teflon-lined autoclave at 150 °C for 15 h. Finally, the resulting product was harvested and named as $RGO/MnFe_2O_4/g\text{-}C_3N_4$.

## RESULTS AND DISCUSSION

The synthesis process of magnetic $RGO/MnFe_2O_4/g\text{-}C_3N_4$ hybrid is illustrated in Fig. 1. We took advantage of the pristine $GO/MnSO_4$ suspension, in which $Mn^{2+}$ ions were combined with the O atoms of the negatively charged surface of GO sheets *via* electrostatic interaction. After adding $FeSO_4$, $g\text{-}C_3N_4$ and KOH, the system was hydrothermally treated at 150 °C for 15 h. In this process, $MnFe_2O_4$ particles were formed ($2Mn^{2+} + 12OH^- + 4Fe^{2+} + O_2 = 2MnFe_2O_4 + 6H_2O$) and expected to be anchored on the RGO surface by Mn–O–H and Fe–O–C bonds. Besides, $MnFe_2O_4$ particles also connected with $g\text{-}C_3N_4$ *via* Fe–N and Mn–N bonds considering the lone pair electrons of N atoms. Furthermore, $g\text{-}C_3N_4$ combined with the remaining –OH groups on the RGO surface to form C–O–C bonds, and the $\pi$-$\pi$ interaction also helped to enhance the combination of $g\text{-}C_3N_4$ and RGO. Hopefully, the as-synthesized $RGO/MnFe_2O_4/g\text{-}C_3N_4$ hybrid with strong interactions will exhibit good catalytic performance for the degradation of MB.

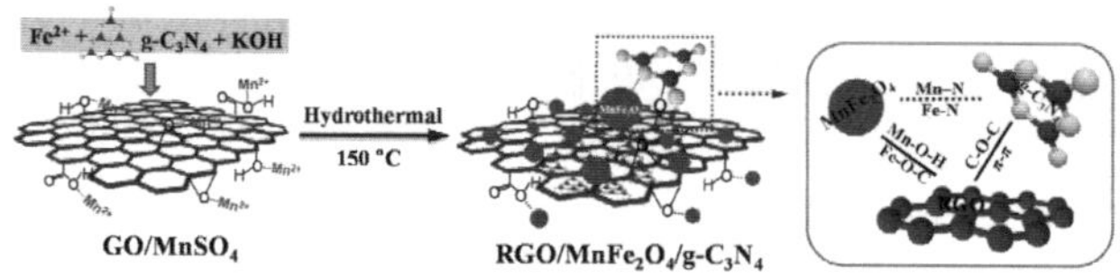


**Fig. 1** Schematic illustration of the synthesis of RGO/$MnFe_2O_4$/g-$C_3N_4$ hybrid and its interfacial interactions.

The MB decolorization of the as-prepared RGO/$MnFe_2O_4$/g-$C_3N_4$ hybrids was examined. As shown in Fig. 2a, it is observed that RGO/$MnFe_2O_4$/g-$C_3N_4$ presents 100% MB decolorization in 80 min under visible light, which is more efficient than those of RGO/$MnFe_2O_4$ (91%), $MnFe_2O_4$ (18%) and pure g-$C_3N_4$ (5.4%). As shown in Fig. 2b, there is no obvious difference in the catalytic performance between RGO/$MnFe_2O_4$/g-$C_3N_4$ and RGO/$MnFe_2O_4$ hybrids without light. The •OH radicals was measured by photoluminescence (PL) technique according to our previous report [4_ENREF_4]. Fig. 2c shows the comparison of PL intensity for different samples. At 20 min, the PL intensities were recorded as the following order: RGO/$MnFe_2O_4$/g-$C_3N_4$> RGO/$MnFe_2O_4$> $MnFe_2O_4$> g-$C_3N_4$, Obviously, RGO/$MnFe_2O_4$/g-$C_3N_4$ hybrid has a significant synergistic effect on the Fenton reaction. As shown in Fig. 2d, the MB decomposition of the catalyst remained 100% in the initial three runs and 99% in the fourth cycle, showing strong catalytic ability in MB decomposition.

251698688

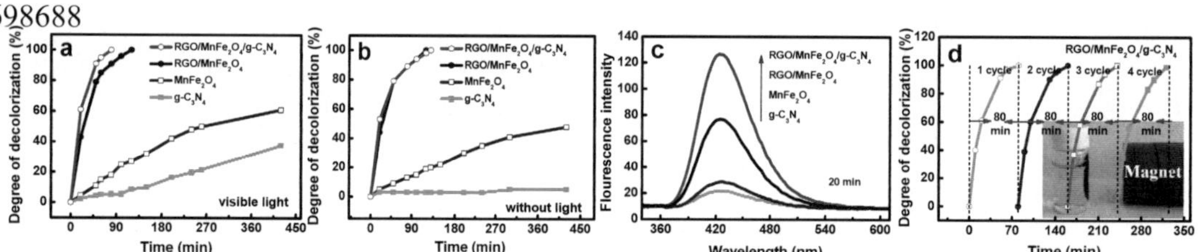


**Fig. 2** Time profiles of MB decolorization with the various samples: (a) under visible light; (b) without light; (c) PL spectra; (d) decomposition of MB with the recycled RGO/$MnFe_2O_4$/g-$C_3N_4$ hybrid (inset is the photo image of the magnetic recovery).

## CONCLUSIONS

In summary, $MnFe_2O_4$ mediated reduced graphene oxide (RGO) and g-$C_3N_4$ sandwich-like hybrids (RGO/$MnFe_2O_4$/g-$C_3N_4$) have been successfully constructed by the hydrothermal method as an effective Fenton-like catalyst. It is found that the formation of C–O–C bonds and π-π interaction between RGO and g-$C_3N_4$ as well as the Fe–N and Mn–N bonds between $MnFe_2O_4$ and g-$C_3N_4$ leads to the strong interactions between RGO, $MnFe_2O_4$ and g-$C_3N_4$ in RGO/$MnFe_2O_4$/g-$C_3N_4$ hybrids. Typically, 50 mL MB (50 mg $L^{-1}$) can be 100% decolorized within 80 min in the presence of $H_2O_2$ at room temperature and the catalyst remains 99% decolorizition in the fourth run.

## REFERENCES


1.Ibrahim, I., Ali, I. O. and Salama, T. M. (2016). "Synthesis of magnetically recyclable spinel ferrite ($MFe_2O_4$, M=Zn, Co, Mn) nanocrystals engineered by sol gel-hydrothermal technology: High catalytic performances for nitroarenes reduction", *Appl. Catal. B: Environ.*, Vol.181, p.389-402.

2.Wu, Z. S., Wang, D. W. and Ren, W. (2010). "Anchoring hydrous $RuO_2$ on graphene sheets for high-performance electrochemical capacitors", *Adv. Funct. Mater.*, Vol.20(20), p.3595-3602.

3.Hou, Y., Wen, Z. and Cui, S. (2013). "Constructing 2D porous graphitic $C_3N_4$ nanosheets/nitrogen-doped graphene/layered $MoS_2$ ternary nanojunction with enhanced photoelectrochemical activity", *Adv. Mater.*, Vol.25(43), p.6291-6297.

4 Peng, X., Qu, J. and Tian S. (2016). "Green fabrication of magnetic recoverable graphene/$MnFe_2O_4$ hybrids for efficient decomposition of methylene blue and the Mn/Fe redox synergetic mechanism", *RSC Adv.*, Vol.6, p. 104549-104555.

# Control of pore structures in MgO-templated carbons for EDLCs

Yuya Kado*, Yasushi Soneda

*Research Institute of Energy Frontier, National Institute of Advanced Industrial Science and Technology, Tsukuba, JAPAN*

*y.kado@aist.go.jp

## INTRODUCTION

Electric double-layer capacitors (EDLCs) have attracted interest for several applications, such as the power-assist systems in electric (hybrid) vehicles due to their high power densities and long life cycles[1,2]. Carbon materials, most commonly, microporous activated carbons (AC) with high specific surface areas are used as the electrode materials in EDLCs. For high power performance, mesoporous carbon materials are more favorable because the ions can migrate more readily through mesopores than micropores, particularly in organic electrolytes[3,4]. We recently reported the positive effect of mesopores on ion diffusion using MgO-templated mesoporous carbons[5,6]. Such MgO-templated mesoporous carbons are attractive because of their mesoporosities, high specific surface areas, and simple production method (e.g., hydrofluoric acid is not required). However, because mesoporous carbons possess large pore volumes, their electrode densities tend to be low. This fact significantly decreases their capacitance per volume. In the present study, we report a simple control of pore structures of the MgO-templated carbon, and investigate the correlation between porous structures of carbons, the electrode densities, and volumetric capacitances in organic electrolytes.

## MATERIALS AND METHODS

Trimagnesium dicitrate nonahydrate (Wako Pure Chemical Industries, Ltd.) was heated to 1000°C in a nitrogen flow (300 mL $min^{-1}$) with ramping at 1, 10, 20, and 40°C $min^{-1}$. After holding at 1000°C for 1 h, the products were cooled to room temperature and immersed in hydrochloric acid to remove any MgO particles. The obtained mesoporous carbons were denoted as MPC-1, MPC-10, MPC-20, and MPC-40, where the numbers correspond to the ramping rate. The pore structures of the carbons were characterized based on their $N_2$ adsorption/desorption isotherms at 77 K.

The active materials were mixed with carbon black and polytetrafluoroethylene (PTFE) in the ratio of 80:10:10 (wt%), and then rolled into 0.1 mm thick films and punched into 10 mm diameter electrodes. The carbon electrodes were dried overnight under a vacuum at 150°C, and the electrode density ($d_{electrode}$) of each was calculated from its weight and thickness. Laminated-type cells in a two-electrode configuration were assembled after the electrodes and separators (TF40-50, Nippon Kodoshi Co.) were immersed in an electrolyte for an hour under reduced pressure. Etched aluminum foils were used as the current collectors. 1 M $TEABF_4$/PC solution was used as electrolyte. Charge/discharge tests were performed in the voltage range of 0–2.5 V at various current densities at room temperature.

## RESULTS AND DISCUSSION

**Figure 1** shows the electrode density, micropore volume, and mesopore volume of the MgO-templated carbons and a commercial microporous activated carbon (AC). It is clear that a faster heating during calcination of magnesium citrate produced a greater pore volume, particularly mesopore volume, in the carbons. The electrode density decreased as the pore volume of the carbons increased. A slower temperature increase provided a higher carbonization yield, suggesting that generated MgO particles were effectively covered with carbon. It is therefore considered that growth of MgO particles as templates was suppressed and small mesopores were formed in the final products. Thus, the pore structures of MgO-templated carbons can be controlled by changing the heating rate during calcination. Volumetric capacitances of these carbons are summarized in **Figure 2**. The carbons obtained at a ramping rate of 1°C $min^{-1}$ exhibited a relatively large volumetric capacitance comparable to that of commercial activated carbons. In addition, they exhibited a better rate capability due to the decreased resistance of ion diffusion by the presence of mesopores.

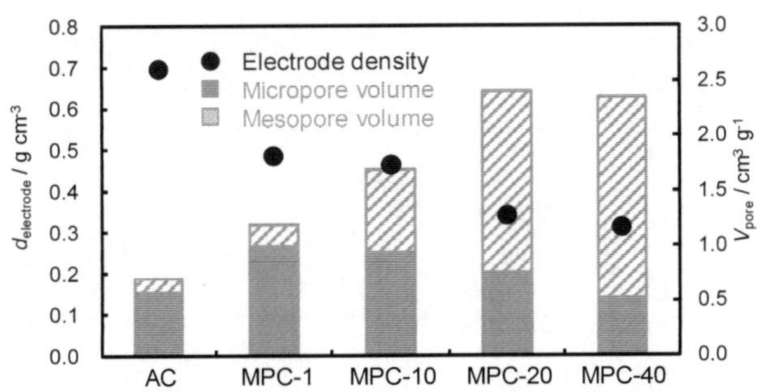


Fig.1 Electrode density plotted with micropore and mesopore volume of the carbons.

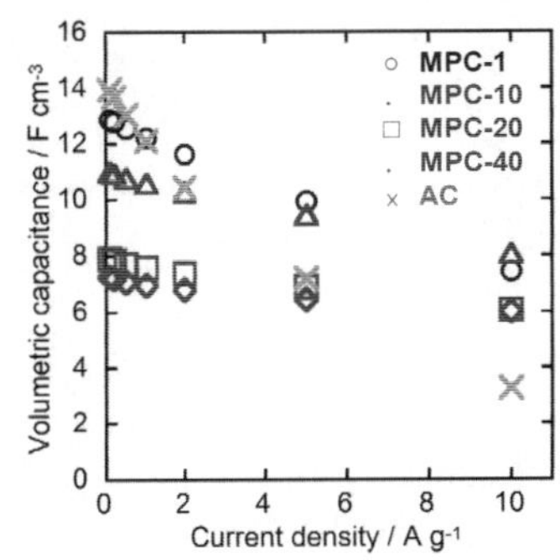


Fig.2 Volumetric capacitance of the carbons.

## CONCLUSIONS

The pore structures of MgO-templated mesoporous carbons can be controlled by changing the heating rate during calcination of magnesium citrate. Clearly, the volumetric capacitance of these mesoporous carbons is higher than that of the commercial microporous activated carbon at high current densities, although the electrode density of the former is inferior to the latter. Thus, the positive effect by existence of mesopores on the rate capability made up for the negative effect of the smaller electrode density.


## ACKNOWLEDGEMENT

A part of this research was performed through collaboration with Toyo Tanso CO. Ltd.


## REFERENCES


1. Conway, B. E. (1999). "Electrochemical Supercapacitors: Scientific Fundamentals and Technological Applications", Kluwer Academic/Plenum Publisher, New York.
2. Simon, P., Gogotsi, Y. (2008). "Materials for electrochemical capacitors", *Nat. Mater.*, Vol. 7, p. 845–854.
3. Inagaki, M. (2009). "Pores in carbon materials-importance of their control", *New Carbon Mater.*, Vol. 24, p. 193–232.
4. Wang, D. W., Li, F., Fang, H. T., Liu, M., Lu, G. Q., Cheng, H. M. (2006). "3D aperiodic hierarchical porous graphitic carbon material for high-rate electrochemical capacitive energy storage", *J. Phys. Chem. B*, Vol. 110, p. 8570–8575.

5. Kado, Y., Imoto, K., Soneda, Y., Yoshizawa N. (2014). “Highly enhanced capacitance of MgO-templated mesoporous carbons in low temperature ionic liquid”, *J. Power Sources*, Vol. 271, p. 377–381.
6. Kado, Y., Soneda, Y., Yoshizawa, N. (2015). “Contribution of mesopores in MgO-templated mesoporous carbons to capacitance in non-aqueous electrolytes”, *J. Power Sources*, Vol. 276, p. 176–180.

# Conversion of Kerosene to Carbon Nanotubes via Co-pyrolysis with Ferrocene

Weerawut Chaiwat[1], Oukrit Thonganatakul[2], Parintorn Vaewhongs[2], Tawatchai Charinpanitkul[2*], Komkrit Suttiponparnit[3]

[1]Environmental Engineering and Disaster Management, School of Interdisciplinary Studies, Kanchanaburi Campus, Mahidol University, Kanchanaburi, Thailand
[2]Center of Excellence in Particle Technology, Department of Chemical Engineering, Faculty of Engineering, Chulalongkorn University, Bangkok, Thailand
[3]Environmental Technology Research Department, PTT Research and Technology Institute, Ayutthaya, Thailand
*ctawat@chula.ac.th

## INTRODUCTION

Kerosene, which is a representative of light distillates fractionated from slop oil, liquid hydrocarbon waste from petroleum refinery, has a potential as carbon source for mass production of carbon nanotubes (CNTs).[1)] Ferrocene could also be employed as additional carbon and Fe catalyst source essential for CNT production. In this study, effect of ferrocene was examined using two portions of ferrocene which were separately supplied into a quartz tube reactor. An appropriate feeding condition of carbon sources and catalysts for synthesis of CNTs via co-pyrolysis at elevated temperatures was experimentally investigated and discussed.

## MATERIALS AND METHODS

### Carbon source and catalyst

As a carbon source, kerosene was provided by PTT (Public) Co., Ltd. Analyses using gas chromatography with mass spectroscopy (GCMS, QP2010SE, Shimadzu) confirmed that the composition in kerosene was the same as light distillates fractioned from slop oil, which is liquid hydrocarbon waste obtained from petroleum refinery. Ferrocene from Sigma-Aldrich was employed as additional sources of carbon and iron catalyst.

### CNT synthesis via co-pyrolysis of kerosene with ferrocene

Two portions of ferrocene were separately supplied into a quartz tube reactor for synthesis of CNTs as shown in **Fig. 1**. The 1$^{st}$ portion of ferrocene in a ceramic boat ($F_1$, 0-20 g) was vaporized by heating up to 550 °C under a nitrogen gas flow of 100 cc/min. The 2$^{nd}$ portion ($F_2$, 10-20 g) was completely mixed with kerosene liquid (125-250 mL) in prior to being fed continuously into the reactor with a flow rate of 3 mL/min. All carbon sources were decomposed and underwent the self-assembly process of CNTs within the reaction zone at elevated temperature in a range of 800-1000 °C. The weight ratio of total ferrocene ($F_1$+$F_2$) and kerosene (K) or the weight ratio of $F_{total}$:K was varied from 1:20 to 1:5. In addition, the weight ratio of ferrocene portions ($F_1$:$F_2$) was also changed for investigating the role of ferrocene in initiating the formation of CNTs.

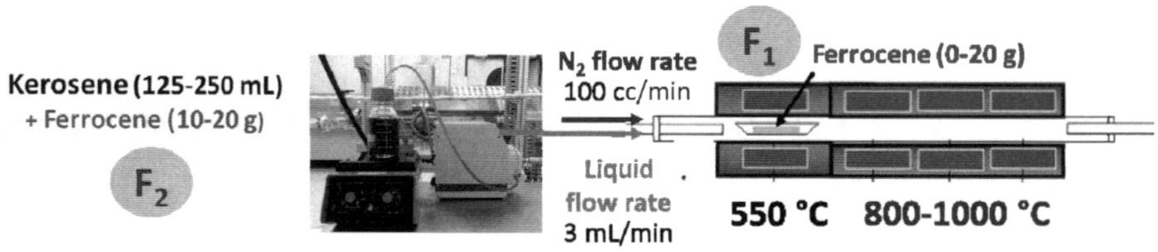


**Fig.1** Schematic diagram of quartz tube reactor system for CNT production from liquid kerosene fed with two different portions of ferrocene catalyst

**RESULTS AND DISCUSSION**

Reaction temperature was varied from 800 to 1000 °C for confirming completed pyrolysis of 125-mL kerosene mixed with ferrocene at a weight ratio of $F_2$:K = 1:10. The pre-fed ferrocene ($F_1$) was also evaporated with a weight ratio of $F_1$: $F_2$ = 1:1. The highest yield of CNTs (over 55 wt%) with more uniform shape and tube diameter less than 100 nm was achieved when synthesizing at 900 °C. The CNTs were rarely found when synthesizing below 850 °C, while the morphology of CNT products was less uniform and the tube diameter was increased when synthesizing above 950 °C. Therefore, 900 °C was selected as a suitable reaction temperature for further investigations. Then twice amount of kerosene at 250 mL was used and the effect of ferrocene fed in two portions was summarized as shown in **Table 1**

**Table 1.** Amount and ratio of ferrocene used as a catalyst and the yield of CNTs produced from 250 mL of kerosene liquid (K) at 900 °C

| Exp. | Ferrocene (F, gram) Pre-fed ferrocene in evap zone ($F_1$) | Ferrocene (F, gram) Ferrocene dissolved in kerosene ($F_2$) | $F_{total}$ | $F_1$:$F_2$ (wt/wt) | $F_2$:K (wt/wt) | $F_{total}$:K (wt/wt) | Yield of CNTs (wt%) |
|---|---|---|---|---|---|---|---|
| (a) | 20 | 20 | 40 | 1:1 | 1:10 | 1:5 | 48.3 |
| (b) | 0 | 20 | 20 | No $F_1$ | 1:10 | 1:10 | 17.6 |
| (c) | 10 | 10 | 20 | 1:1 | 1:20 | 1:10 | 26.7 |
| (d) | 0 | 10 | 10 | No $F_1$ | 1:20 | 1:20 | 22.7 |

At a constant temperature of 900 °C regardless of weight ratio of kerosene and ferrocene as summarized in **Table 1**, SEM analyses of the synthesized products revealed bundles of long and curly CNTs with a diameter of 20-100 nm as shown in **Fig. 2**. However, the highest CNT yield (almost 50 wt%) could be achieved when using the lower weight ratio of ferrocene to kerosene ($F_2$:K) with the presence of $F_1$, as also summarized in **Table 1**. It is interesting that the 1$^{st}$ portion of ferrocene put in the evaporation zone ($F_1$), which was pre-fed into the system to initiate the CNT products, could provide a significant contribution to an increase in CNT yields.

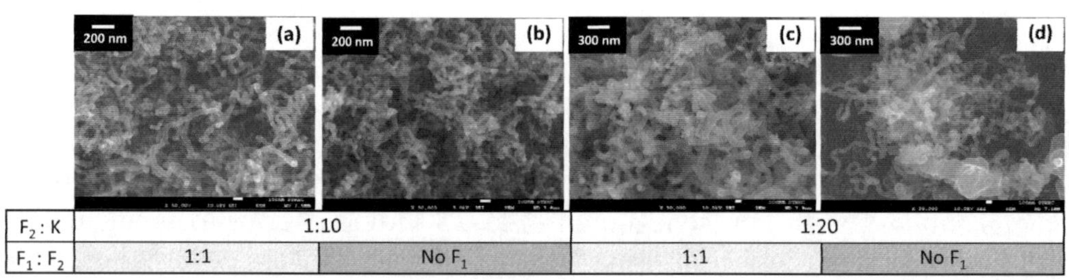


**Fig.2** SEM images of CNTs synthesized at 900 °C from ferrocene-dissolved kerosene with $F_2$:K (wt/wt) at (a-b) 1:10 and (c-d) 1:20 with and without pre-fed ferrocene ($F_1$)

**CONCLUSIONS**

Kerosene as a representative hydrocarbon from petroleum waste could be converted to CNTs with the presence of ferrocene. Bundles of long and curly CNTs with a tube diameter of 20-100 nm could be suitably synthesized at 900 °C via co-pyrolysis with two portions of ferrocene catalyst. The 1$^{st}$

portion of ferrocene ($F_1$), which was pre-fed into the quartz tube reactor system to initiate the CNT products, could exert significant contribution to an increase in CNT yields up to nearly 50 wt%.

## REFERENCES

1. Kerdnawee, K., Termvidchakorn, C., Yaisanga, P., Pakchamsai, J., Chookiat, C., Eiad-ua, A., Wongwiriyapan, W., Chaiwat, W., Ratchahat, S., Faungnawakij, K., Suttiponparnit, K., Charinpanitkul, T. (2017). "Present Advancement of Carbonaceous Nanomaterial Production from Industrial Waste and Promising Applications," *Kona Powder Part J*, Vol.34, p.24-43 (2017).

# COR and OER of Several Carbon Materials in KOH Solution

Taro Kinumoto*, Makoto Eto, Kohei Ono, Miki Matsuoka,
Tomoki Tsumura, Masahiro Toyoda

Division of Applied Chemistry, Department of Integrated Science and Technology,
Faculty of Science and Technology, Oita University
700 Dannoharu, Oita, 870-1192, Japan
*kinumoto@oita-u.ac.jp

## INTRODUCTION

Electrochemical power sources that use oxygen as a reactant may play an essential role in the future energy needs of the society. In particular, fuel cells, aqueous metal-air batteries, and metal hydride (MH)/air batteries hold promise as power sources beyond lithium-ion batteries (LIBs). For such electrochemical power sources, an alkaline solution is often employed as the electrolyte and an oxygen electrode reaction (ORR; oxygen reduction reaction and OER; oxygen evolution reaction) occurs at the cathode. For instance, if a MH is used as the anode (in the case of a MH/air battery), the anode and cathode half reactions and the overall reaction are described as follows:

Anode $4MH + 4OH^- \leftrightarrows 4M + 4H_2O + 4e^-$ [1]

Cathode $O_2 + 2H_2O + 4e^- \leftrightarrows 4OH^-$ [2]

Total $O_2 + 4\,MH \leftrightarrows 4\,M + 2H_2O$ $E_{emf} = 1.229$ V [3]

The reaction kinetics is slow at the cathode and therefore, an electrode catalyst is required. A perovskite-type metal oxide such as $LaMnO_3$ is one of the catalyst candidates; however, the electric conductivity is not sufficient for practical applications. Thereby, the cathode is often constructed with carbon materials such as carbon black, activated carbon, or graphite powders to support the electric conductivity. Unfortunately, the carbon material can be oxidized by a reaction called carbon oxidation reaction (COR), as shown in reaction [4]; this phenomenon is of concern during the charge reaction where the electrode potential exceeds 1.229 V.

$C + 6OH^- \rightarrow CO_3^{2-} + 3H_2O + 4e^-$ $E_0 = 0.27$ V *vs.* NHE [4]

In this presentation, the COR and OER behavior of some carbon materials such as Ketjen black (KB), natural graphite (NG) and vapor grown carbon fiber (VGCF) in KOH aqueous solution are reported [1].

## MATERIALS AND METHODS

### Material

Ketjen black (KB, EC-300J, LION Co. Ltd., Japan), natural graphite (NG, SMO-3M, SEC carbon Co. Ltd., Japan) and vapor grown carbon fiber (VGCF, Showa denko Co. Ltd.,) were used without any pre-treatment.

### Electrochemical test using rotation ring-disk electrode technique

COR and OER behavior of some carbon materials were investigated by using a rotating ring-disk electrode (RRDE) technique. The disk electrode was made of glassy carbon (GC, Tokai Carbon, diameter = 5 mm). The ink prepared with the carbon materials, ethanol and water. Then, the inks were applied dropwise to GC disk. After that, an anion-exchange-membrane ionomer solution (AS-4, TOKUYAMA Corp.) was applied to a calculated thickness of 80 nm. The ring electrode was made of platinum to detect oxygen and $HO_2^-$ that were produced at the disk electrode during the electrochemical measurements.
A perfluoroalkoxyalkane (PFA) beaker (250 mL) and 0.1 M KOH solution was used as the test cell and the electrolyte, respectively. Hg|HgO reference electrode was placed at the vicinity of the ring-disk working electrode. Ni mesh was placed on the bottom and was used as the counter electrode.
The hydrodynamic voltammetry was carried out at 0.001 mV/sec. An electrode rotator (AFMSRCE, Pine Research Instrumentation) was used and the rotation speed of the RRDE was set at 1600 rpm.

## RESULTS AND DISCUSSION

**Fig. 1** shows the hydrodynamic voltammograms for the carbon material and the ring current at -0.5 V for ORR. The disk current is shown as the specific current per the electric capacitance. In the case of KB, the disc current was obvious over 0.23 V; on the other hand, the current responsible for ORR was recorded more positive than 0.63 V. The latter fact suggests that the onset potential of OER is 0.63 V because ORR should be only taken place when oxygen is produced at the disk electrode. Therefore, the oxidation current that was recorded from 0.23 V should be ascribed to the COR of KB according to [4]. Nevertheless, both of OER and COR were taken place when the potential was more positive than 0.63 V. From the results, we can consider the potential dependency for behaviour of the carbon materials. The results are summarized in **Fig. 2**. The kinetically negligible region and the COR region can be defined whether COR did not take place or not. The competition reaction region means that the potential range in where OER and COR were taken place simultaneously.

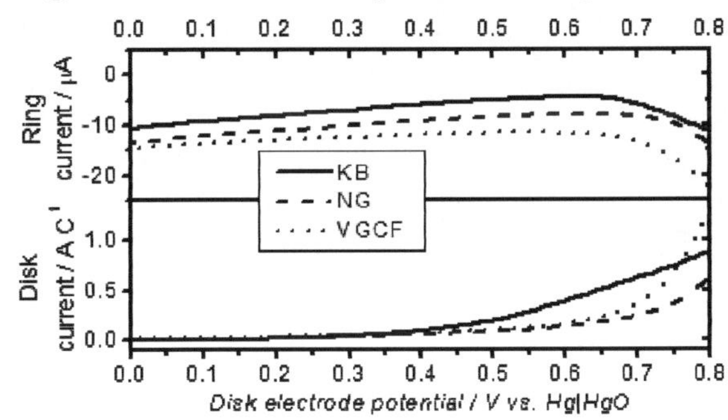


**Fig. 1** a) hydrodynamic voltammograms for disk electrode of KB, NG and VGCF, b) ring current at -0.5 V for ORR.

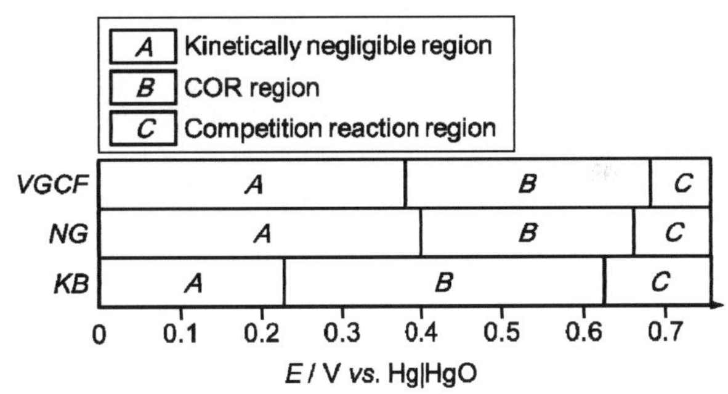


**Fig. 2** Schematic diagram proposed for COR behavior of KB, NG and VGCF over 0 V *vs*. Hg|HgO in 0.1 M KOH aq.

## CONCLUSIONS

We investigated the COR and OER of KB, NG and VGCF in KOH solution. We could determine that the onset potential for COR for each carbon materials such as 0.23 V of KB, 0.40 V of NG and 0.38 V of VGCF. The results of this study give guidance for carbonaceous materials to use in the positive electrode of the rechargeable air battery where an alkaline medium is employed.

## ACKNOWLEDGEMENT

This work was supported by the "Development of Hydrogen/Air Secondary Battery, Advanced Low Carbon Technology Research and Development Program (ALCA)" of the Japan Science and Technology Agency (JST).

## REFERENCES

1. T. Kinumoto, M. Eto, K. Ono, M. Matsuoka, T. Tsumura and M. Toyoda, TANSO, No. 273, pp. 101-106 (2016) [in Japanese].

# Defect Engineering of Graphene and Other 2D Materials

Ruitao Lv[1*], Mingxiang Hu[1], Feiyu Kang[1], Mauricio Terrones[2]

[1] School of Materials Science and Engineering, Tsinghua University, Beijing, 100084, China
[2] Department of Physics, The Pennsylvania State University, University Park, PA 16802, USA
*lvruitao@tsinghua.edu.cn

## INTRODUCTION

Defects are usually regarded as imperfections in materials that could significantly degrade their performance. However, at the nanoscale, defects (e.g. dopants, vacancies) can be exploited to generate novel and useful materials and devices. Therefore, defect engineering of 2D materials is becoming an important research line so far. In particular, substitutional doping with heteroatoms is a very efficient way to tailor the electrical and chemical properties of graphene. Both experimental and theoretical research on N-doped graphene have been extensively carried out. However, the experimental reports on large-area Si-doped (SiG) and B-doped graphene (BG) sheets are still quite scarce so far. In this contribution, we summarize our recent progress on controllable synthesis of centimeter size, high-crystallinity doped graphene sheets[1, 2]. Their excellent sensing properties are demonstrated by using gaseous (e.g. $NO_2$, $NH_3$) and liquid molecules (e.g. crystal violet, rhodamine B and methylene blue) as probes[1, 2]. Besides graphene, other 2D layered materials (e.g. transition metal dichalcogenides, TMDs) are attracting worldwide research attentions[3]. We also demonstrated the defect engineering of TMDs for ultrasensitive pressure sensing[4] and highly efficient hydrogen evolution reaction (HER) catalysis[5].

## MATERIALS AND METHODS

The synthesis of large-area SiG and BG was achieved in a home-designed bubbler-assisted ambient-pressure chemical vapor deposition (AP-CVD) system operating at ambient pressure. Methoxytrimethylsilane (MTMS, $C_4H_{12}OSi$, CAS No.: 1825-61-2, Aldrich) and hexane ($C_6H_{14}$, CAS No.: 110-54-3, Purity>99.0%, Sigma-Aldrich) were used as Si and C precursors, respectively. MTMS can also serve as C precursor in addition to hexane in the SiG growth. More experimental details can be found elsewhere [1, 2, 4, 5].

## RESULTS AND DISCUSSION

Large-area SiG sheets were synthesized using MTMS and hexane as silicon and carbon precursors, respectively. The size and shape of the resulting SiG sheets can be well tuned by a conformal growth on Cu foils. Si-doped graphene can efficiently quench the fluorescence and thus enhance the Raman scattering of organic molecules (Fig.1a-b). The theoretical modeling of SiG indicates that the substitution of C by Si induces local curvature around the substitutional site, which also enhances the interaction of the tested dye molecules with graphene (Fig.1c-d). The interacting orbitals of $RhB^+$ were shown to spread due to the increased interaction with Si-doped sites, thus providing a more effective mechanism of charge transfer between dye and substrate and an overall enhanced GERS signals. This strategy can also be used to grow large-area and high-quality BG sheets, as shown in Fig.1e-f. A peculiar feature detailed in Fig.1g is consistently observed throughout the sample. Considering the length scale and its symmetry, it is inferred that the defect should be more complex than a simple B substitution. *Ab-initio* STM simulations provide the $B_3$-dopants with the same symmetry (see inset of Fig.1h). The feature at -0.41 V in the experimental scanning tunneling spectroscopy (STS) shown in Fig.1h can be attributed to the $B_3$-dopants. In order to investigate the effect of B doping on graphene's gas sensitivity, comparative studies on $NO_2$ and $NH_3$ under similar detection environment were carried out. It is found that B doping of graphene has enhanced its sensitivity 27 and 105 times to reach detection limits of 95 ppt and 60 ppb (s/n=3), respectively, for $NO_2$ and $NH_3$ detections.

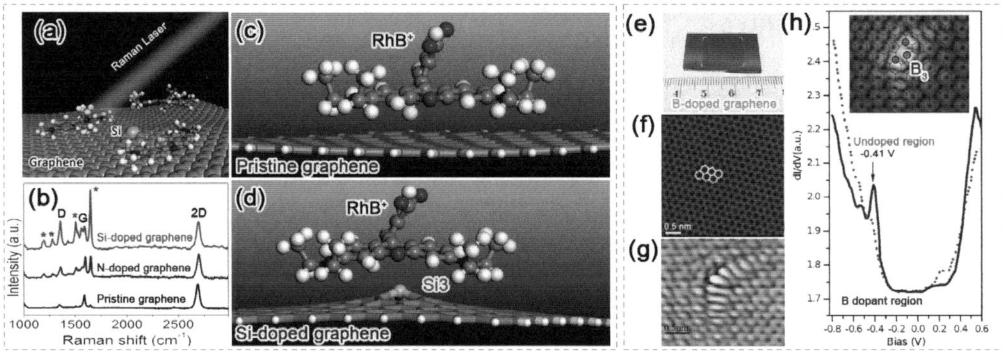


**Fig.1** Large-area heteroatom-doped graphene sheets. (a) Schematic illustration of SiG and different probe molecules. (b) Raman signals of RhB molecules on pristine, N-doped and Si-doped graphene sheets. Optimized geometries of $RhB^+$ adsorbed on (c) pristine and (d) Si-doped graphene sheets. (e-f) Morphology of BG sheets. (h) *dI/dV* curves measured on $B_3$-dopants (blue solid curve) and on undoped graphene region (red dot curve). Inset is the simulated STM image and corresponding ball-stick structural model of $B_3$-dopant.

Large-area few-layer $MoS_2$ films with different thickness were synthesized by metal e-beam evaporation followed by ambient-pressure sulfurization. As-synthesized $MoS_2$ films are uniform and homogeneous. High-performance F-P sensors were then assembled by transferring as-synthesized few-layer $MoS_2$ films onto the end face of a ceramic ferrule with a sensitivity of 89.3 nm/Pa[4], which is 3 orders of magnitude higher than those of conventional diaphragm materials (e.g. silica, silver films). Moreover, by introducing sulfur vacancies in the atomic thin $WS_2$ sheets, free-standing and flexible $WS_2/WO_{2.9}/C$ hybrids on graphite foils can be directly employed as the electrode for electrocatalytic HER in acidic solution, which demonstrate outstanding performance of a low Tafel slope of 36 mV $dec^{-1}$, an onset potential of -20 mV[5]. Such electrocatalytic performance is superior to most of the results reported for TMDs so far.

## CONCLUSIONS

In conclusion, we have proposed a versatile and controllable strategy to synthesize large-area heteroatom-doped graphene sheets using liquid precursors. As-synthesized SiG can efficiently quench the fluorescence and thus enhance the Raman scattering of organic molecules. Unique croissant-like features corresponding to $B_3$-dopants embedded within the hexagonal BG lattice are visualized for the first time with STM/STS techniques. As-synthesized BG can be applied to detect gas molecules at extremely low concentrations. Compared with pristine graphene, boron doping of graphene can significantly enhance the sensitivity 27 and 105 times for $NO_2$ and $NH_3$ detections, respectively. By rational defect design, TMDs can be engineered to be used for pressure sensing and highly efficient HER catalysts. Our results will open up new avenues for developing high-performance sensors and generating clean energy with 2D materials.

## REFERENCES


1. Lv, R., et al. (2014). "Large-Area Si-Doped Graphene: Controllable Synthesis and Enhanced Molecular Sensing", *Adv. Mater.*, Vol. 26, p.7593-7599 (Back Cover)

2. Lv, R., et al. (2015). "Ultrasensitive gas detection of large-area boron-doped graphene", *PNAS*, Vol. 112, p.14527-14532

3. Lv, R., et al. (2015). "Transition Metal Dichalcogenides and Beyond: Synthesis, Properties, and Applications of Single- and Few-Layer Nanosheets", *Acc. Chem. Res.*, Vol. 48, p.56-64

4. Lv, R.*, et al. (2017). "Ultrasensitive Pressure Detection of Few-Layer $MoS_2$", *Adv. Mater.*, Vol. 29, p.1603266 (Inside Cover)

5. Lv, R.*, et al. (2017). "Noble-Metal-Free Hybrid Membranes for Highly Efficient Hydrogen Evolution", *Adv. Mater.*, Vol. 29, p.1603617

# Development of Carbon Spheres with Tailored Porosity, Structure and Functionality

Mietek Jaroniec*

Department of Chemistry and Biochemistry, Kent State University
Kent, OH 44242 USA
*jaroniec@kent.edu

A significant advancement has been made in the development of carbons and carbon-based composites with well-defined porosity, which is well illustrated by a vast number of reports devoted to ordered mesoporous carbons discovered in 1999 [1]. Also, extensive efforts have been made toward controlling morphology of carbon materials, especially carbon spheres (for instance, see recent review in Nature Materials [2]). The aim of this presentation is to provide a critical appraisal of recent progress and prospects in the development of carbon spheres (CSs) with tailored sizes, porosity, morphology and functionality for specific applications ranging from adsorption of pollutants from air and water, adsorption of biomolecules, heterogeneous catalysis to drug delivery, energy storage and conversion. Some guidelines for the design and synthesis of CSs will be provided; namely, (i) how to control the size and distribution of CSs, (ii) how to fabricate the entirely microporous, mesoporous and micro-mesoporous CSs with desired distribution, orientation and volume of pores, (iii) how to obtain CSs with incorporated inorganic nanoparticles, (iv) how to fabricate hollow, core-shell and yolk-shell CSs with various structures, composition and functionality, and (v) how to design composite spheres consisting of amorphous or graphitic carbon, graphene and inorganic nanoparticles/nanosheets.

In particular, numerous works have been published on the synthesis of CSs by using the modified Stöber method [3], which opens new opportunities for the development of CSs with tailored porosity and doped species, and/or incorporated or in-situ formed inorganic nanoparticles [4,5]. This method is well suited for the synthesis of CSs with narrow particle size distribution and tailorable particle size ranging from nanosized to micron size particles. Besides normal porous CSs, more complex structures such as hollow CSs, core-shell and yolk-shell structures as well as carbon-inorganic composite structures can be prepared by adjusting experimental conditions and using suitable precursors, additives and catalysts.

Figure 1 illustrates tremendous opportunities of the Stöber method for the preparation of various types of CSs and composite spheres. For instance, normal microporous CSs can be prepared by carbonization of phenolic resin spheres formed by $NH_3$-catalyzed polymerization of resorcinol and formaldehyde in ethanol-water solution, however N-doped CSs are obtained when ethylenediamine (EDA) is used instead of $NH_3$ [6]; in this case, EDA acts as catalyst and N-source. Depending on the type of metal, addition of inorganic salts during Stöber synthesis can be used to incorporate metal or metal oxide nanoparticles into CSs as well as activate these spheres to develop additional microporosity [7,8]. For instance, the use of Ni, Co or Fe salts leads to graphitic CSs with incorporated metal nanoparticles (NPs), while K or Na salts can act as in-situ activators, which leads to highly microporous CSs. Also, inorganic NPs as colloidal silica can be used in this synthesis to obtain silica-carbon composite spheres, which after dissolving silica are transformed to CSs with uniform spherical mesopores. However, the controlled use of tetraethylorthosilicate (TEOS) during Stöber synthesis allows one to obtain a variety of structures, depending on the amount of TEOS used, ranging from normal silica-carbon composite spheres to core-shell and yolk-shell complex structures, which after silica dissolution can give hollow CSs (HCSs) or core-shell CSs, and after carbon burn off the aforementioned composite spheres can be transformed to silica spheres [9].

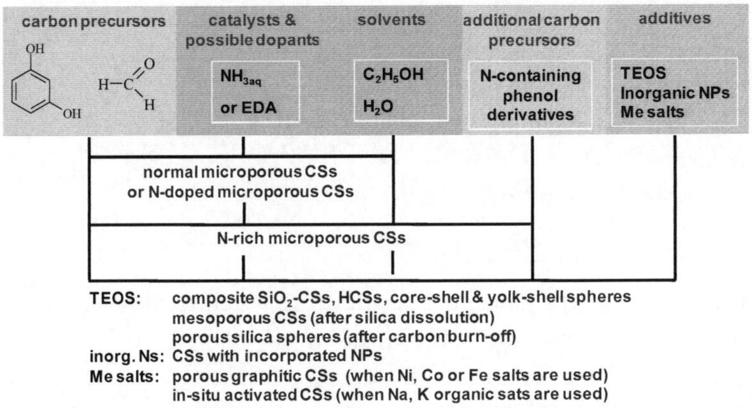


**Figure 1**. Illustration of synthetic opportunities of Stöber method for preparation of various carbon-containing composite spheres and CSs with tailored porosity, composition and structures.

## REFERENCES


1. Ryoo, R., Joo, S. H., Jun, S. (1999). "Synthesis of highly ordered carbon molecular sieves via template-mediated structural transformation", *J. Phys. Chem. B* Vol.103 (37), p.7743–7746
1. Liu, J., Wickramaratne, N. P., Qiao, S. Z., Jaroniec, M. (2015). "Molecular-based design and emerging applications of nanoporous carbon spheres", *Nature Mater.* Vol.14(8), p.763−774
2. Liu, J., Qiao, S. Z., Liu, H., Chen, J., Orpe, A., Zhao, D. Y., Lu, G. Q. (2011) "Extension of the Stöber method to the preparation of monodisperse resorcinol formaldehyde resin polymer and carbon spheres", *Angew. Chem., Int. Ed.* Vol.50(26), p.5947−5951
3. Choma, J., Jamioła, D., Augustynek, K., Marszewski, M., Gao, M., Jaroniec, M. (2012) "New opportunities in Stöber synthesis: preparation of microporous and mesoporous carbon spheres", *J. Mater. Chem.* Vol.22(25). p.12636-12642
4. Tian, H., Liu, J., O'Donnell, K., Liu, T., Liu, X., Yan, Z., Liu, S., Jaroniec, M. (2016) "Revisiting the St□ber method: Design of nitrogen-doped porous carbon spheres from molecular precursors of different chemical structures", *J. Coll. Interface Sci.* Vol.476, p.55-61
5. Wickramaratne, N. P., Xu, J., Wang, M., Zhu, L., Dai, L., Jaroniec, M. (2014) "Nitrogen enriched porous carbon spheres: Attractive materials for supercapacitor electrodes and $CO_2$ adsorption", *Chem. Mater.* Vol.26(9), p.2820-28281
6. Wickramaratne, N.P., Jaroniec, M. (2014) "Facile synthesis of polymer and carbon spheres decorated with highly dispersed metal nanoparticles", *Chem. Commun.* Vol. 50(82), p.12341-12343
7. Ludwinowicz, J., Jaroniec, M. (2015) "Potassium salt-assisted synthesis of highly microporous carbon spheres for $CO_2$ adsorption", *Carbon* Vol.82, p.297−303

# Dielectric Behavior of Carbon Materials

D.D.L. Chung[1*]
[1]Composite Materials Research Laboratory,
Department of Mechanical and Aerospace Engineering, University at Buffalo,
The State University of New York, Buffalo, NY 14260-4400, U.S.A.
[*]ddlchung@buffalo.edu

## INTRODUCTION

The electrical behavior of a material pertains to the conduction behavior (related to the electrical conductivity, which governs the electrical resistance) and the dielectric behavior (related to the electric permittivity, which governs the capacitance, and to the effectiveness for electromagnetic interference shielding, or EMI shielding). Both magnetic and dielectric interactions are involved in EMI shielding, so the EMI shielding effectiveness does not simply relate to the dielectric behavior. Furthermore, shielding typically involves frequencies $\geq$ 1 GHz, which are much higher than the frequencies (typically $\leq$ 1 kHz) that pertain to electrochemical applications. Concerning carbon materials, much attention has been given to the conductive behavior, but little attention has been given to the dielectric behavior, except for work related to EMI shielding. Relevant applications pertaining to the dielectric behavior include electrochemical electrodes, electrically conductive additives, fuel-cell bipolar plates, EMI shielding, structural capacitors and capacitance-based structural self-sensing (nondestructive evaluation, or NDE). This paper reviews the science and applications of the dielectric behavior of carbon materials, with emphasis on frequencies $\leq$ 1 kHz.

## DIELECTRIC BEHAVIOR OF CARBON MATERIALS

The relative permittivity of carbon materials varies over a wide range, due to the differences in structure and surface functional groups. For example, its value at 50 Hz ranges from 31 for carbon black [1] to 124 for activated carbon [1], to 915 for graphite oxide [2,3], and to 1130 for reduced graphite oxide [4].

Carbon black is widely used as an electrically conductive additive in nonconductive electrochemical electrodes such as manganese dioxide ($MnO_2$). Carbon black deforms greatly upon compression between the stiff $MnO_2$ particles. The deformation enhances the dielectric connectivity, thereby affecting the permittivity of the carbon black [5]. The presence of a liquid electrolyte also enhances the dielectric connectivity and hence the permittivity of the carbon black [5].

## DIELECTRIC BEHAVIOR OF CARBON FIBER POLYMER-MATRIX STRUCTURAL COMPOSITES

Continuous carbon fiber polymer-matrix composites are lightweight structural materials. Without modification of the composite, the high fringing-field-induced capacitance, as measured by using electrodes that are separated from the composite by an insulating film, is sensitive to the damage in the composite, thereby providing a new method of NDE for carbon materials, as shown for a continuous carbon fiber thermosplastic-matrix composite [6]. Moreover, the through-thickness permittivity can be tailored by modifying the interlaminar interface, thereby obtaining positive and negative capacitors, the series combination (antiferroelectric coupling) of which gives relative permittivity as high as 78,000 [7].

## COMMON PITFALLS IN DIELECTRIC RESEARCH

Pitfalls are common in dielectric research. A pitfall due to inadequate recognition of the limitation of an impedance meter (which is not meant for measuring the capacitance of a conductor) has resulted in incorrectly huge values of the permittivity in the literature [8,9]. An electrically insulating film needs to be positioned between the specimen and electrode.

The widely used technique of impedance spectroscopy measures the impedance as a function of frequency and uses the frequency dependence to extract information. The frequency dependence is conventionally described in terms of the Nyquist plot, which is mathematical fitted through modelling with an equivalent electrical circuit that is intended to describe the electrical/dielectric behaviour of the material. The circuit model obtained by the curve fitting tends to be not unique, so the determined values of the circuit elements in the model are not very meaningful.

In the parallel-plate capacitor geometry, a specimen is sandwiched by two electrical contacts. Both the specimen and each of the two specimen-contact interfaces contribute to the measured capacitance. Thus, the two contributions should be decoupled. This decoupling can be achieved by simply measuring the capacitance for three or more thicknesses of the specimen material [1]. However, the decoupling is commonly not conducted and the contribution from the interfaces is commonly neglected.

## CONCLUSION

The dielectric behavior of carbon materials is rich in science and applications, though it has not received wide attention. Carbon materials studied include particulate carbons, carbon-$MnO_2$ particulate mixtures, and continuous carbon fiber polymer-matrix composites.

## REFERENCES


1. Wang, A. and Chung, D. D. L. (2014). "Dielectric and electrical conduction behavior of carbon paste electrochemical electrodes, with decoupling of carbon, electrolyte and interface contributions", *Carbon*, Vol.72, p.135-151
1. Hong, X., Yu, W., Wang, A. and Chung, D. D. L. (2016). "Graphite oxide paper as a polarizable electrical conductor in the through-thickness direction", *Carbon*, Vol.109, p.874-882
2. Hong, X., Yu, W. and Chung, D. D. L. (2017). "Significant effect of sorbed water on the electrical and dielectric behavior of graphite oxide", *Carbon*, Vol.119, p.403-418
3. Hong, X., Yu, W. and Chung, D. D. L. (2017). Electric permittivity of reduced graphite oxide. *Carbon*, Vol.111, p.182-190
4. Moalleminejad, M. and Chung, D. D. L. (2015). "Dielectric constant and electrical conductivity of carbon black as an electrically conductive additive in a manganese-dioxide electrochemical electrode, and their dependence on electrolyte permeation", *Carbon*, Vol.91, p.76-87
5. Chung, D. D. L. and Eddib, A. A. "Capacitance-based nondestructive evaluation of carbon fiber polymer-matrix composite", *Carbon*, to be submitted.
6. Takizawa, Y. and Chung, D. D. L. (2016). "Continuous carbon fiber polymer-matrix composites in unprecedented antiferroelectric coupling providing exceptionally high through-thickness electric permittivity", *J. Mater. Sci.*, Vol.51(14), p.6913-6932
7. Sarkara, S., Mondala, A., Deyb, K. and Raya, R. (2016) "Defect driven tailoring of colossal dielectricity of Reduced Graphene Oxide", *Mater. Res. Bull.*, Vol.74, p.465–471
8. Sachdev, V. K., Sharma, S. K., Bhattacharya, S., Patel, K., Mehra, N. C., Gupta, V. and Tandon, R. P. (2015). "Electromagnetic shielding performance of graphite in cement matrix for applied application", *Adv. Mater. Lett.*, Vol.6(11), p.965-972

# Doped Carbon Materials as Electrocatalysts for the Oxygen Reduction Reaction

M.J. Nieto-Monge[1], J.C. Ruiz-Cornejo[1], D. Sebastián[1], C. Alegre[1,2], M.V. Martínez-Huerta[3], E. Pastor[4], M.J. Lázaro[1*]

[1]Instituto de Carboquímica (CSIC), Miguel Luesma Castán 4, 50018 Zaragoza, Spain
[2]Istituto di Tecnologie Avanzate per l'Energia "Nicola Giordano" (CNR), Via Salita S. Lucia sopra Contesse 5, 98126 Messina, Italy
[3]Instituto de Catálisis y Petroleoquímica (CSIC), Marie Curie 2, 28049 Madrid, Spain.
[4]Instituto de Materiales y Nanotecnología. Universidad de La Laguna. Ap. postal 456. 38206 La Laguna, Tenerife, Spain
*mlazaro@icb.csic.es

## INTRODUCTION

Nowadays, carbon materials are used as catalyst support in different energy conversion devices, such as polymer electrolyte membrane fuel cells (PEMFCs). Doping carbon materials with heteroatoms like nitrogen or sulphur appears as a promising strategy to improve the behaviour as catalyst support. Moreover, recent works have also pointed out a significant electroactivity for the oxygen reduction reaction (ORR) of nitrogen-doped carbon materials without the noble metal active phase [1], which implies a breakthrough towards cost-efficient fuel cell systems. In this work, carbon nanostructures are chemically treated to introduce nitrogen or sulphur functionalities within the carbon surface/matrix. Ethylenediamine and urea have been studied as precursors of nitrogen while elemental sulphur is investigated as precursor of sulphur species. The right choice of treatment conditions and precursor nature relies on the carbon material characteristics. The catalytic activity of these doped carbon materials towards the ORR is electrochemically evaluated in alkaline environment and discussed with regard to the incorporation of N and S heteroatoms.

## MATERIALS AND METHODS

Carbon nanofibers (CNFs) have been synthesized by the decomposition of methane on a Ni-based catalyst at either 600ºC (CNF600) or 700ºC (CNF700). The resulting CNFs have been treated in acid to remove the traces of Ni used for the growth of CNFs and to partially functionalize the surface. Ethylenediamine or urea have been mixed with CNF700, followed by a thermal treatment procedure, resulting in CNF700-E and CNF700-U samples. Elemental sulphur has been mixed with CNF600 and CNF700 and treated at 160ºC to obtain S-doped CNFs (CNF600-S and CNF700-S).The electrocatalytic activity towards the ORR has been assessed in a three-electrode electrochemical cell. CNFs were deposited on a glassy carbon based rotating disk electrode (RDE) with a loading of 500 μg $cm^{-2}$. The base electrolyte was a 0.1 M NaOH aqueous solution, a reversible hydrogen electrode (RHE) was used as reference, and a high area glassy carbon rod was employed as counter electrode.

## RESULTS AND DISCUSSION

Table 1 summarizes the chemical composition of doped carbon nanofibers as determined by elemental analysis. The procedures used for nitrogen introduction lead to a content of 0.18 wt% and

0.38 wt% for the methods using urea and ethylenediamine, respectively. These relatively low percentages might be associated with the low surface area of CNF700 (72 $m^2$ $g^{-1}$) compared to CNF600 (140 $m^2$ $g^{-1}$). Whereas, CNFs treated with sulphur present 2.6 wt% and 4.9 wt% S, indicating a good yield of the process.

**Table 1. Chemical composition of the samples determined by elemental analysis.**

| Sample | C | H | N | S | O |
|---|---|---|---|---|---|
| CNF700 | 97.2 | 0.08 | 0 | 0 | 0.34 |
| CNF700-E | 97.4 | 0.15 | 0.38 | 0 | - |
| CNF700-U | 97.4 | 0.11 | 0.18 | 0 | - |
| CNF700-S | 94.7 | 0.09 | 0 | 2.6 | 0.40 |
| CNF600 | 96.9 | 0.26 | 0 | 0 | 0.86 |
| CNF600-S | 92.5 | 0.26 | 0 | 4.9 | 1.05 |

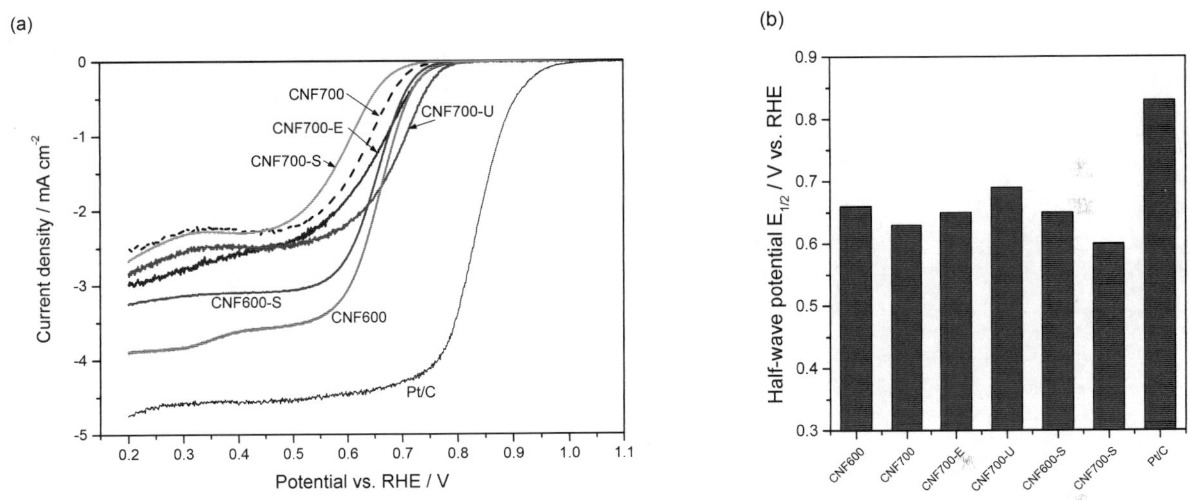


**Fig. 1. (a) Linear sweep voltammetries (LSVs) recorded in RDE in $O_2$-saturated 0.1 M NaOH solution at 1600rpm. (b) Half-wave potential from LSV curves.**

Figure 1 shows the linear sweep voltammetry curves related to the investigation of ORR in alkaline electrolyte. All curves display the typical wave-like shape with a current plateau when scanning from 1 V vs. RHE towards more negative potentials. The bar chart plot summarizes the values of half-wave potential for each sample. The results are still far from a benchmark Pt/C catalyst but CNFs exhibit some electroactivity towards the ORR. Doping with sulphur does not appear to improve the activity, while doping with nitrogen does, even for nitrogen content below 1 wt%.

## CONCLUSIONS

Chemical treatments of CNFs are effective in introducing low content of nitrogen or sulphur. The presence of nitrogen appears to be beneficial in the electrochemical activity of CNFs towards the ORR in alkaline environment.

## ACKNOWLEDGMENTS

The authors gratefully acknowledge the financial support given by the Spanish Ministry of Economy and Competitiveness (MINECO) through Projects ENE2014-52158-C2-1-R and 2-R. M.J.N.M. thanks MINECO for her FPI grant.

**REFERENCES**

1. Ratso, S., Kruusenberg, I., Vikkisk, M., Joost, U., Shulga, E., Kink, I., Kallio, T. & Tammeveski, K. Highly active nitrogen-doped few-layer graphene/carbon nanotube composite electrocatalyst for oxygen reduction reaction in alkaline media. *Carbon N. Y.* **73,** 361–370 (2014).

# Effect of Ball-milling on Supercapacitive Performance of Expanded Graphite

Yue Dong[1], Su Zhang[2], Huaihe Song[1*], Xiaohong Chen[1], Jisheng Zhou[1], Zhaokun Ma[1], Dianzeng Jia[2]

[1]State Key Laboratory of Chemical Resource Engineering, Beijing University of Chemistry Technology, Beijing, P. R. China
[2]Key Laboratory of Energy Materials Chemistry Ministry of Education, Xinjiang University, Urumqi, P. R. China.
*songhh@mail.buct.edu.cn

## INTRODUCTION

Supercapacitors is a promising candidate for the next generation of energy storage devices due to their high power density, long life span, low maintenance cost and etc. However, most researches focused on the gravimetric performance and overlooked the volumetric properties, which is more valuable in practical applications. Herein, we prepared a dense material from expanded graphite by a facile ball-milling process and performed as electrodes for symmetric supercapacitors. Both the gravimetric and volumetric performance were tested.

## MATERIALS AND METHODS

### Material

The graphite oxide was prepared by modified Hummers' method. Typically, natural graphite was used as raw material and a mixture of $KMnO_4$ and $H_2SO_4$ was the oxidant. The graphite oxide was broke up into powder and exfoliated in a muffle oven at 1273K for 30 s to get the expanded graphite nanosheets (GNS). The ball-milling process was then performed in a high energy mill for different time. The products were named as GNS-x, in which "x" represents the time of ball-milling.

## RESULTS AND DISCUSSION

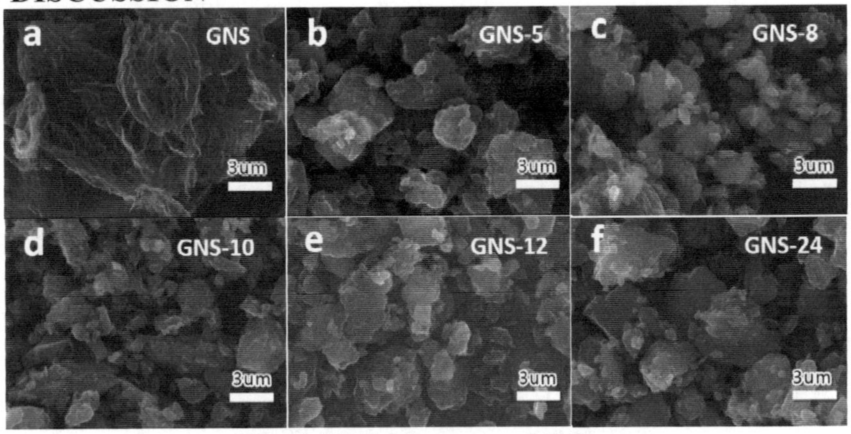


**Fig. 1. SEM images of EG-x.**

Fig. 1. shows the SEM images of GNS-x. The GNS have loose and porous structure consisting of a large amount thin flakes. After ball-milling, the GNS-x become a dense and bulk material. Besides, the size of GNS-8 is smallest, which means the GNS-8 may has most edge structure.

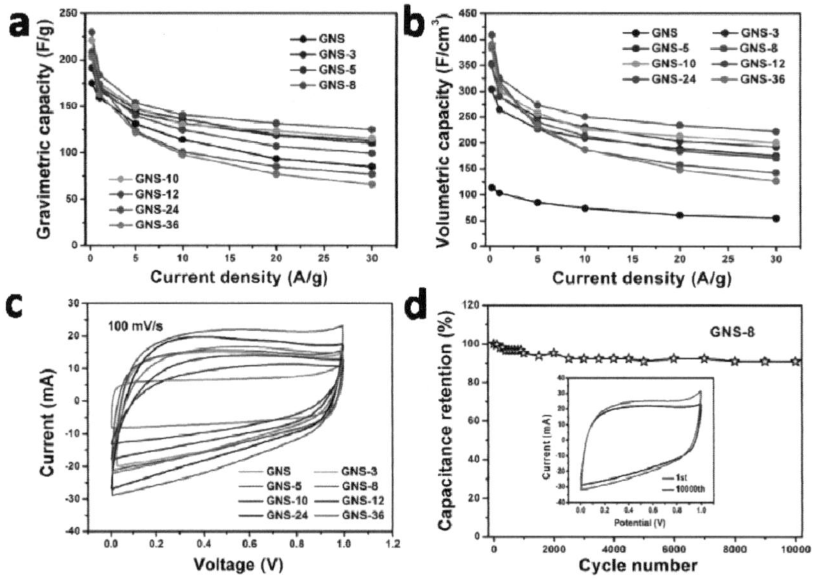


**Fig. 2. gravimetric (a) and volumetric capacitance (b) of GNS and GNS-x as a function of current density, CV curves of GNS-x at scan rate of 100mV/s (c), cycling stability of the GNS-8 at a current density of 10 A/g, the inset shows CV curves before and after the 10000th cycles at 100 mV/s (d).**

The specific area and pore volume decrease drastically after ball-milling (typically, the specific area decrease from 248.8 m$^2$ g$^{-1}$ of GNS to 29.7 m$^2$ g$^{-1}$ of GNS-8, and pore volume decrease from 1.03 cm$^3$ g$^{-1}$ of GNS to 0.06 cm$^3$ g$^{-1}$ of GNS-8), but both the gravimetric and volumetric are enhanced (Fig. 2a and 2b). The CV curves are nearly rectangle, which indicates the absolute double layer capacitance behaviour (Fig. 2c). After 10000 cycling at current density of 10 A g$^{-1}$, the capacity retention of GNS-8 is 91%, and the CV curves are similar before and after 10000$^{th}$ cycles (Fig. 2d), which demonstrates the excellent stability and super reversibility. Besides, we find the content of heteroatoms increase after ball-milling through element analysis, which would provide pseudocapacitance for supercapacitors.

## CONCLUSIONS

From the discussion above, we discovered that both the gravimetric and volumetric were enhanced even though the specific area and pore volume decreased massively. We primarily deduced that the reason of this unusual result might be the synergistic effects of the defects and pseudocapacitance of heteroatoms. Our work could help for the development and optimization of active materials for high performance supercapacitors.

## REFERENCES

1. Choi, W. S., Shim, W. G. and Ryu, D. W. (2012). "Effect of ball milling on electrochemical characteristics of walnut shell-based carbon electrodes for EDLCs", *Microporous and Mesoporous Mater.*, Vol.155(1), p.274-280

2. Zhang, C., Lv, W. and Yang, Q. H. (2015). "Towards superior volumetric performance design and preparation of novel carbon materials for energy storage", *Energy Environ. Sci.*, Vol.8(5), p.1390-1404

# Effect of Carbon Structure on Ru Catalyst for Ammonia Synthesis

Masayasu Nishi[1], Saori Taira[2], Hideyuki Takagi[1]*

[1]Research Institute of Energy Frontier (RIEF), National Institute of Advanced Industrial Science and Technology (AIST), Tsukuba, Ibaraki 305-8569, Japan
[2]Tsukuba University, Tsukuba, Ibaraki 305-8577, Japan
*hide-takagi@aist.go.jp

## INTRODUCTION

Ammonia ($NH_3$) that can be synthesized by $CO_2$-free ammonia synthesis process using supported Ru catalysts under mild condition has been widely studied as an energy carrier for hydrogen storage, transportation and utilization in the past decade. Although several supported Ru catalysts had been developed for synthesis of $NH_3$ at lower pressure and temperature[1], the activity of supported Ru catalysts in synthesis of $NH_3$ is strongly related to their structural property, carbon supports especially, but their correlation is still unclear. Our recent studies exhibited that mesoporous carbon (MPC)-supported Ru catalysts were highly active in synthesis of $NH_3$ under mild condition (< 400 °C and 1 MPa), presumably associated with the structural property of MPC supports. Herein, a series of MPC-supported Ru catalysts with different structural properties and Ru loadings was prepared, and their activity in synthesis of $NH_3$ under mild condition was thoroughly elucidated.

## MATERIALS AND METHODS

### Material

Three MPC supports (product code: P(3)010) were purchased from Toyo Tanso Co., Ltd., Japan, and pre-treated at 1500, 1800, and 2100 °C (designated as MPC-1500, MPC-1800 and MPC-2100, respectively) under an inert atmosphere before used. Activated carbon (AC, product code: HG15-119) as a reference support was purchased from Osaka Gas Chemical Co., Ltd., Japan. The structural property of these carbon supports, i.e. surface area ($S_{BET}$) and pore size (P) is listed in Table 1, based on the $N_2$ physisorption data recorded at 77 K. The supported Ru catalysts with a Ru loading of 2.5 or 10 wt%, and a proper amount of Cs promoter were prepared by impregnation method, designated as *x*Ru/*y* (*x*: the Ru amount (wt%), *y*: carbon support).

**Table 1.** Structural property of carbons.

| Carbon | $S_{BET}$ ($m^2\ g^{-1}$) | P (nm) |
|---|---|---|
| MPC-1500 | 1180 | 10 |
| MPC-1800 | 910 | 11 |
| MPC-2100 | 280 | 14 |
| AC | 1700 | 0.9 |

### Method

Synthesis of $NH_3$ over supported Ru catalysts was performed in a fixed-bed reaction system using a space velocity (SV) of 9000 $h^{-1}$, and a $H_2/N_2$ volume ratio of 3 at 350-400 °C and 0.99 MPa. At specific reaction periods, the gas products were *in situ* analysed by a Shimadzu GC-2014 instrument equipped a TCD detector.

## RESULTS AND DISCUSSION

Fig.1 shows the activity of supported Ru catalysts with various carbon supports and Ru loadings in synthesis of $NH_3$ at 350 °C and 0.99 MPa. It is obvious that the supported Ru catalysts using MPC-series carbon supports gave higher activity than that using AC support, probably due to that the molecular diffusion through mesoporous structures can be facilitated. On the other hand, the activity of supported Ru catalysts in synthesis of $NH_3$ can be largely improved if the Ru loading is increased to 10 wt%. Among them, the 10Ru/MPC-1800 catalyst gave the highest activity in synthesis of $NH_3$ under mild condition.

Fig. 2 shows the high-resolution transmission electron microscope (HRTEM) images of supported Ru catalysts using MPC-series carbon supports. The Ru particle sizes of 10Ru/MPC-1500 and 10Ru/MPC-1800 with high $S_{BET}$ values are around 2 nm while those of 10Ru/MPC-2100 with a low $S_{BET}$ value is increased to 5 nm. Clearly, Ru can be finely dispersed on the MPC supports of high surface areas, and the corresponding Ru catalysts with more Ru active sites give high activity in synthesis of $NH_3$ under mild condition. By contrast, the 10Ru/MPC-2100 catalyst with relatively small $S_{BET}$ value and large Ru particle sizes showed relatively low activity in synthesis of $NH_3$. It is also noticed that the crystallinity of MPC support of 10Ru/MPC-1800 is slightly higher than that of 10Ru/MPC-1500. It is presumably that relatively high activity of 10Ru/MPC-1800 than that of 10Ru/MPC-1500 in synthesis of $NH_3$ should be also associated with higher crystallinity of MPC support accompanying with good electronic property.

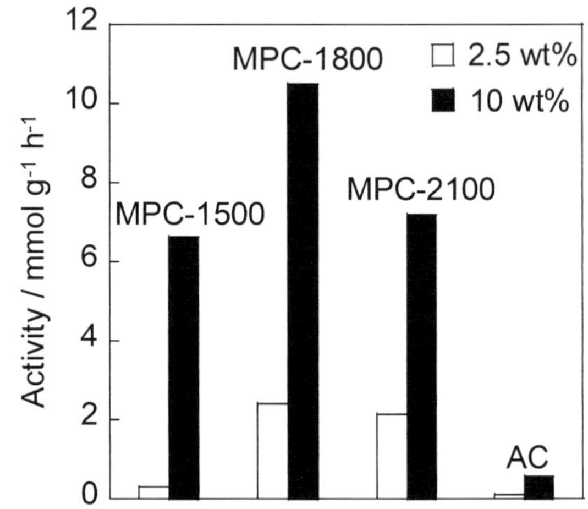


**Fig. 1.** Influences of carbon supports and Ru loadings on activity of supported Ru catalysts in synthesis of $NH_3$.

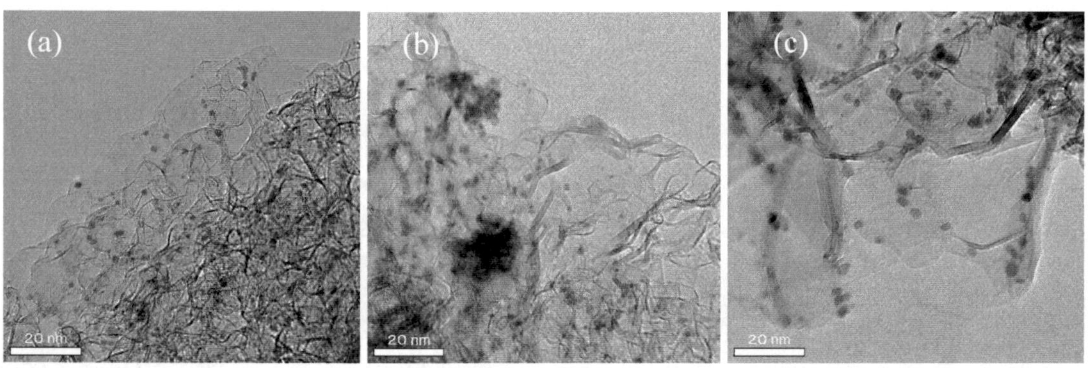


**Fig. 2.** HRTEM images of (a) 10Ru/MPC-1500, (b) 10Ru/MPC-1800 and (c) 10Ru/MPC-2100.

## CONCLUSIONS

Influences of carbon supports and Ru loadings on activity of supported Ru catalysts using MPC and AC as carbon supports with different pre-treatment in synthesis of $NH_3$ under mild condition were discussed. Among them, the 10Ru/MPC-1800 catalyst with suitable $S_{BET}$ value ($m^2$ $g^{-1}$), Ru particle size (2 nm) and semi-crystalline carbon framework gave the highest activity in synthesis of $NH_3$ at 350-400 °C and 0.99 MPa using a SV of 9000 $h^{-1}$, and a $H_2/N_2$ volume ratio of 3.


## ACKNOWLEDGEMENT

This work was supported by the Council for Science, Technology and Innovation (CSTI), Cross-ministerial Strategic Innovation Promotion Program (SIP), "energy carrier" (funding agency: JST).


## REFERENCES

1. Aika, K., Hori, H. and Ozaki, A. (1972). "Activation of Nitrogen by Alkali Metal Promoted Transition Metal I. Ammonia Synthesis over Ruthenium Promoted by Alkali Metal", *J. Cat.*, Vol.27(8), p.424-431.

# Electrochemical oxidation of carbon xerogels and correlation with physico-chemical properties

David Sebastián[1], Cinthia Alegre[1], María E. Gálvez[2], María J. Lázaro[1*]

[1]Instituto de Carboquímica, CSIC, Zaragoza, Spain
[2]University of Sorbonne, UPMC Paris, Jean le Rond D'Alembert Institute, UMR CNRS, Saint-Cyr L'Ecole, France
*mlazaro@icb.csic.es

## INTRODUCTION

Carbon xerogels (CXGs) are of great interest from a quality-to-cost point of view among the different gel-derived carbon materials (mainly comprising also aerogels and cryogels). Advanced carbon-based materials are playing a key role nowadays in the development of low-cost and environmentally friendly electrochemical devices devoted to the conversion and storage of energy, such as fuel cells, electrolyzers, batteries, solar cells and supercapacitors. Regardless the more or less graphitic nature of carbon materials, in the presence of electron acceptor species like water, the occurrence of relatively high positive potentials (> 1 V vs. RHE) unavoidably results in a certain degree of carbon oxidation reaction (COR), and thus, the degradation and eventual modification of the electrode initial properties. The discussion about the electro-oxidation mechanism and the influence of carbon characteristics is still an open question, even though it is known that a high graphitic character and a low surface area help to reduce the oxidation rate. CXGs, though possessing a certain ordering degree in the short-range, are not graphitic materials. This fact, together with their developed porosity, make them more prone to corrosion processes than graphite-like carbons. The main scope of the present work is to investigate the electro-oxidation behaviour in acidic solution of CXGs differing in porosity and surface chemistry. The control of some specific CXGs features like the surface chemistry and carbonization conditions appears of paramount importance to design strategies to minimize degradation phenomena related to carbon electro-oxidation.

## MATERIALS AND METHODS

Carbon xerogels with different physico-chemical properties were synthesized according to the sol-gel methodology[1]. In order to modify textural and structural properties, three synthesis variables were selected: the resorcinol to sodium carbonate molar ratio (800 and 1050), the pyrolysis temperature (800 and 900 °C) and the solvent used during polymerization of the gel (water and a mixture water/acetone). Briefly, resorcinol (1,3-dihydroxybenzoic acid, Sigma-Aldrich), sodium carbonate (Panreac), formaldehyde (Sigma-Aldrich) and the solvent (water or acetone/water 1:9 w/w) were thoroughly mixed together in a closed vessel, and the pH was adjusted to 6.0 with diluted nitric acid. All carbon xerogels were synthesized with stoichiometric resorcinol to formaldehyde ratio (1:2) and a dilution ratio of 5.7 (i.e. moles of solid per mole of solvent). The polymerization and curing processes took place in the closed vessels containing the mixture for one day at room temperature, one day at 50 °C and five days at 85 °C, as described in [2]. Afterwards, the so-obtained organic gels were submerged in acetone for 3 days, replacing the liquid with fresh acetone every day. The drying process was performed in sub-critical conditions: 5 h at 65°C and 5 h at 110°C. The resulting powder was milled and pyrolyzed at 800 °C or 900 °C for 3 h under inert ($N_2$) atmosphere. Carbon xerogels were named as follows: CXG, for carbon xerogel, 8 or 10 for the resorcinol-to-sodium carbonate ratio, A for the mixture of water/acetone and 8 or 9 depending on the pyrolysis temperature.

## RESULTS AND DISCUSSION

The carbon oxidation reaction was studied under potentiostatic conditions (by holding a constant potential of 1.2 V vs. RHE in acid electrolyte and monitoring the variation of current with time) in a three-electrode electrochemical cell. The COR was investigated, according to the Kinoshita-Giordano methodology [3,4]. The current associated to carbon oxidation is shown in Fig.1. The current-time curves follow the typical Kinoshita-Giordano trend following $kt^{-n}$ like behaviour, with $k$ and $n$ being constants. CXG8-8, with the largest surface area, presents the highest oxidation rate. Besides, the amount of oxygen and the pyrolysis temperature play a role in hindering corrosion. CXGs characterized by a low amount of O (CXG-10-8 and CXG-8-9) are less prone to corrosion. A higher pyrolysis temperature leads to a higher ordering degree of the carbon material, making it more resistant to oxidation.

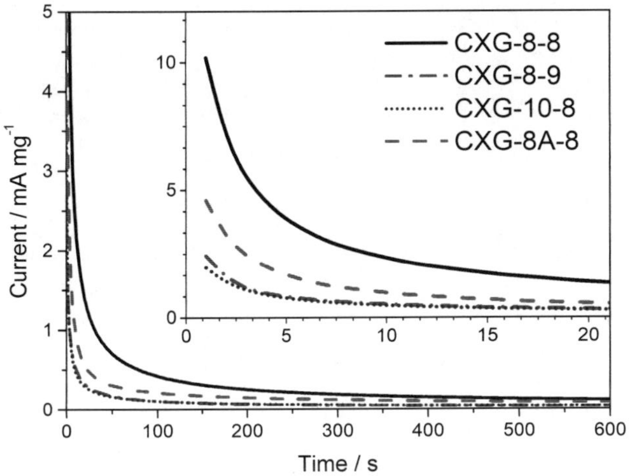


**Fig. 1. Current-time behaviour of carbon xerogels at 1.2 V vs. RHE in deaerated 0.5 M $H_2SO_4$, gas diffusion electrode with 0.6 mg cm$^{-2}$ loading and 33 wt% Nafion as ionomer.**

| Sample | $S_{BET}$ m$^2$ g$^{-1}$ | $V_{pore}$ cm$^3$ g$^{-1}$ | O1s % |
|---|---|---|---|
| CXG8-8 | 536 | 1.80 | 3.3 |
| CXG8-9 | 508 | 1.66 | 1.5 |
| CXG-10-8 | 440 | 0.45 | 4.4 |
| CXG-8A-8 | 425 | 0.39 | 5.7 |

**Table 1. Surface area (BET) and pore volume determined from $N_2$ physisorption isotherms (-196 °C) and oxygen content from X-ray photoelectron spectroscopy.**

## CONCLUSIONS

Selecting the synthesis conditions of carbon xerogels is extremely important in order to design materials with a high resistance to oxidation. Porosity features and pyrolysis temperature appear necessary to improve the resistance of carbon materials towards electrochemical oxidation.

## REFERENCES


1. Pekala, R. W. Organic aerogels from the polycondensation of resorcinol with formaldehyde. *J. Mater. Sci.* **24,** 3221–3227 (1989).
2. Maldonado-Hódar, F. J. J., Ferro-García, M. A., Rivera-Utrilla, J., Moreno-Castilla, C., Ferro-García, M. A., Rivera-Utrilla, J. & Moreno-Castilla, C. Synthesis and textural characteristics of organic aerogels, transition-metal-containing organic aerogels and their carbonized derivatives. *Carbon N. Y.* **37,** 1199–1205 (1999).
3. Kinoshita, K. & Bett, J. Electrochemical oxidation of carbon black in concentrated phosphoric acid at 135°C. *Carbon N. Y.* **11,** 237–247 (1973).

4. Giordano, N., Antonucci, P. L., Passalacqua, E., Pino, L., Aricò, A. S. & Kinoshita, K. Relationship between physicochemical properties and electrooxidation behaviour of carbon materials. *Electrochim. Acta* **36,** 1931–1935 (1991).

# Electrolytic Treatment Of Dry-jet Wet Spun Carbon Fibers

Ma, Y [1], Liu, J[1], Xue, Y[1], Liang, J[1], Wang, X[1]

[1] State Key Laboratory of Organic-Inorganic Composites, Beijing University of Chemical Technology, Chao-Yang District, Beijing 100029, China Flinders University,
my132593@126.com

## INTRODUCTION

Smooth surface and small diameters gave dry-jet wet spun carbon fibers excellent tensile strength. However, the smooth surface also result in weak interface properties between carbon fibers and epoxy resin. In this study, we performed electrolytic treatment to adjust the surface properties of dry-jet wet spun carbon fibers, and explored the impact of electrolyzing time on the mechanical properties of carbon fibers.

## MATERIALS AND METHODS

### Material

Un-sized and un-treated PAN-based CFs with 12000 filaments per tow were supplied by the Zhongfu shenying Carbon Fiber Co., Ltd. (Jiangsu, China). The matrix resin selected in this study was the Epoxy 6101 resin purchased from the Nantong Xingchen Synthetic Material Co., Ltd. (Nantong, China), and the selected curing agent was triethylenetetramine purchased from the Beijing Chemical Reagents Co. (Beijing, China).

### Electrochemical oxidation treatment

Ammonium bicarbonate purchased from the Beijing Chemical Reagents Co. (Beijing, China) was studied to prepare the aqueous solutions for the electrochemical oxidation treatment of the CFs. The ammonium bicarbonate concentration was 1M and the electric current was 1000mA. During the treatment, the treatment temperature was 30 degrees Celsius□the time periods for the treatment were set at 90, 120, 150 s. After the treatment, the samples were rinsed with deionized water and then dried at 100□. A schematic representation of the treatment is shown in Fig. 1.

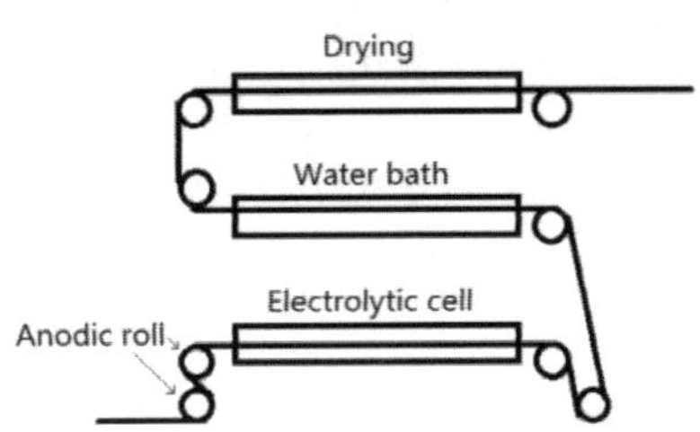


**Fig.1. Schematic representation of the electrochemical oxidation treatment.**

### Evaluations of tensile strength, elastic modulus, and interlaminar shear strength

The mechanical properties of the carbon fiber multifilament spline were tested using the Instumb5567 model universal testing machine (US). During the tensile strength measurements, the sample was fixed to the fixture, and effective test length was 150mm. The movement speed of the fixture was set to 10mm / min. While for measuring the ILSS, we adopted a standard procedure of GB3357-82. The composites contained 40% CFs and 60% resin by volume. Each type of CFs was first wound onto a rectangular framework in parallel; then the aligned fibers were impregnated into the Epoxy 6101 resin mixed with curing agent (triethylenetetramine). The mass ratio of the resin and the curing agent was set at 10/1. Then, the samples were cured in an oven at120 ◦C for 2 h.

## RESULTS AND DISCUSSION

The mechanical performance test is shown in **Fig2**. Before 90s, the tensile strength(TS) decreased with time, while the ILSS did not change significantly. After 90 s, the tensile strength is opposite to the interlayer shear strength. We can see that in the experimental test range, the tensile strength reached the maximum at 120s, while the ILSS showed the lowest value. This may be due to the change of nitrogen content in the fibers. The specific values of TS and ILSS are shown in **Table 1**.

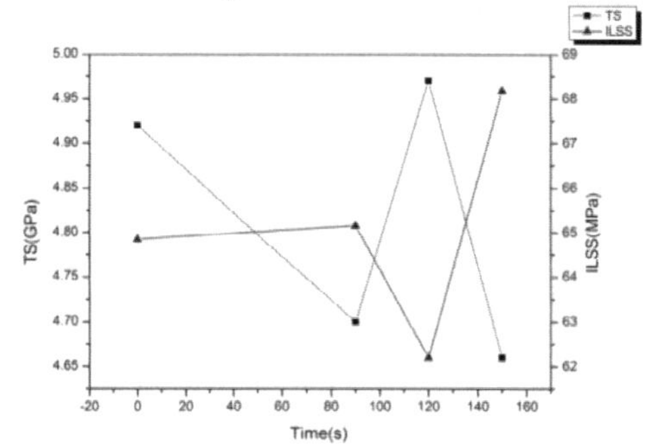


**Fig. 2. Dependence on time of the mechanical properties of CFs treated with $NH_4HCO_3$**

Because of the smooth surface and small diameters, dry-jet wet spun carbon fibers show the obvious surface inertia which mainly shows the combination of functional groups between fibers and resin in this experiment. Therefore, the interlaminar shear strength of these composites is lower than that of the wet spun carbon fiber composites, so even if the time reaches 150s, ILSS does not show a high degree of improvement.

**Table1. The specific values of tensile strength, Young's modulus and ILSS**

| | **YM(Gpa)** | **CV(%)** | **TS(Gpa)** | **CV(%)** | **ILSS(Mpa)** | **CV(%)** |
|---|---|---|---|---|---|---|
| 0s | 242.23 | 2.21 | 4.92 | 4.21 | 64.86 | 1.33 |
| 90s | 247.62 | 1.67 | 4.7 | 4.45 | 65.16 | 0.35 |
| 120s | 234.61 | 1.34 | 4.97 | 5.17 | 62.2 | 0.81 |
| 150s | 235.86 | 2.23 | 4.66 | 4.52 | 68.17 | 2.07 |
| T700 | 226.08 | 2.97 | 4.62 | 2.78 | 61.33 | 1.17 |

## CONCLUSIONS

The surface of dry-jet wet spun carbon fibers is inert, so the resulting ILSS of is lower than that of wet spun carbon fiber. When we choose 1M $NH_4HCO_3$ as the electrolyte, and select 1000mA for the current treatment, no obvious improvement of ILSS of dry-jet wet spun carbon fiber/epoxy composite was observed by changing the time before 150s. To show more obvious changes, we may need to choose a higher current or longer time.

## REFERENCES

1. Jie Liu, Yuli Tian, Yujia Chen, et al. A surface treatment technique of electrochemical oxidation to simultaneously improve the interfacial bonding strength and the tensile strength of PAN-based carbon fibers [J]. Materials Chemistry and Physics,2010,122 :548–555.

# Enhancement of Mechanical Properties of CFRTP Using Polymer Colloids

Katsumasa Uematsu, Tetsuya Yamamoto*, Toshihira Iirisawa, Yasuhiro Tanabe

Graduate School of Engineering, Nagoya University,Nagoya, Japan
*yamamoto.tetsuya@material.nagoya-u.ac.jp

## INTRODUCTION

Carbon fiber reinforced thermoplastic(CFRTP) has recently attracted attention mainly due to their potential of recyclability, high production rate, and low cost compared with conventional CFRP. The surface adhesion between the resin and the carbon fiber in CFRTP is one of the most important factors for enhancement of mechanical properties. However, because the thermoplastic resin shows low interfacial adhesion with carbon fiber, weak strength of the material is a big problem for practical use. In addition, because the thermoplastic resin has a high melt viscosity, it is difficult to impregnate of resin between carbon fibers when molding. A molded product with insufficient impregnation causes a decrease in mechanical properties. The research on improvement of interfacial adhesion with a thermoplastic resin was carried out by various methods. For example, modification with a silane coupling agent on the surface of a carbon fiber has been reported. However, it is not so much as to correspond to the interfacial adhesion force between the thermosetting resin and the carbon fiber. There is no report that impregnation is improved by these methods. In the present study, the purpose is to improve the adhesion between the thermoplastic resin and the carbon fiber and the impregnating property. The method is adsorbing the polymer fine particles on carbon fibers using the electrophoresis technique performed in polymer colloids.

## MATERIALS AND METHODS

### Materials

The polymer colloids were prepared as follows: The powders of Polyamide 12 (PA12, SP-500, TORAY, Japan), surfactant and electrolyte were added to water with ultrasonic operation for 30 seconds. Polyamide 6 (PA6, CM1001, TORAY, Japan) were used as the matrix resin, and PAN-based carbon fibers (HTS40, Toho Tenax, Japan) plain fabrics were used as reinforcement fibers.

### Test methods

The carbon fibers were immersed in the polymer colloid to allow the polymeric particles to adsorb on their surfaces with the help of an electrophoresis system, as shown in Fig. 1. The adsorption time was set to 30 second. The morphologies of the carbon fibers with particles were observed by field emission scanning electron microcopy (FE-SEM, JSM-7500FA, JEOL, JAPAN). To calculate the surface coverage of the PA 12 particles on the carbon fibers, a thermal gravimetric analyzer (TG, DTG-60AH, SHIMADZU, Japan) was used. The interfacial shear strength (IFSS) between a carbon fiber and PA6 film was measured by using fragmentation methods. For evaluation of impregnating property, the impregnation ratio was calculated from the observation of the cross-sctional area of CFRTP with FE-SEM.

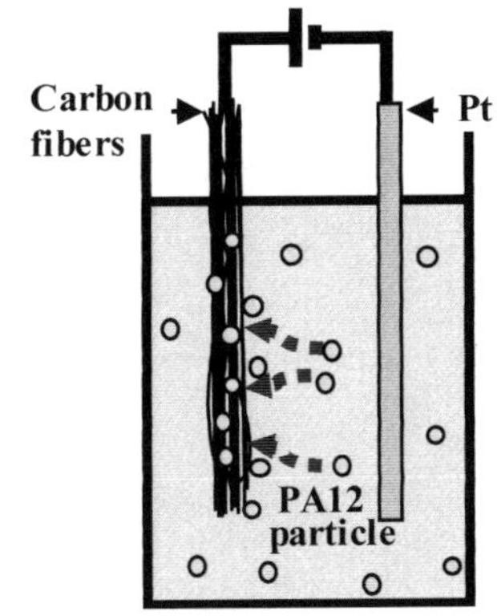


**Fig.1** System of PA12 particles adsorbed on carbon fiber using nylon colloid and electrophoresis.

## RESULTS AND DISCUSSION

Fig. 2 shows the SEM image of a modified carbon fiber at the applied voltage 30 V. The coverage of the PA 12 particles on the carbon fibers was calculated to be 0.45. In contrast, surface coverage was 0.07 when carbon fibers were coated with PA 12 powders under the atmosphere. Herein, the colloid and electrophosis technique was usueful for the surface-modification.

The IFSS between a carbon fiber and PA6 film is shown in Fig.3. When PA12 particles were adsorbed on carbon fibers, the IFSS improved by 54.3% compared with carbon fibers without particles. It was possible to think that the particles played the role of adhesive. Our previous work reported that IFSS improved when the coverage ratio of polymer particles was high [1]. Because there were some spaces to be coverd with PA12 particles, further improvement of IFSS can be expected with the size of PA12 reduced.

Fig.4 shows SEM images of cross section of CFRTP. There were many voids in CFRTP without PA12 particles. On the other hand, there are few voids in CFRTP with PA12 particles. Adsorption of particles of the resin component on the carbon fiber bundle caused the void to be buried and made it easier to impregnate the carbon fiber bundle with the base resin component. The impregnation ratio was 92.3% without particles, while it was 96.3% using PA12 particles. Thus, it was possible to make CFRTP without voids by the particle adsorption technique.

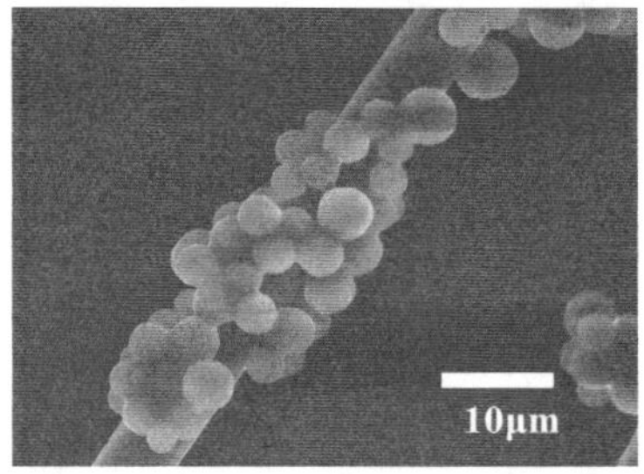


**Fig.2** SEM image of the carbon fibers modified by PA12 colloids at the applied voltage of 30V.

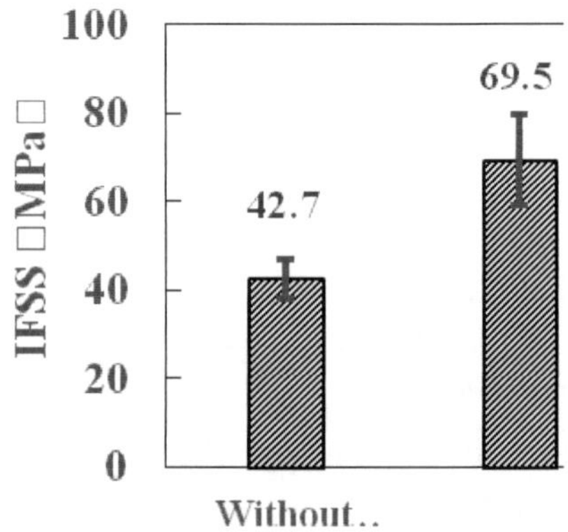


**Fig.3** Effect of PA12 particles adsorbed onto the carbon fiber on the IFSS between carbon fiber and PA6 film.

## REFERENCES

1. T.Yamamoto, K. Uematsu, T. Irisawa and Y. Tanabe , *Composites Part A*, 88, 75-78 (2016)

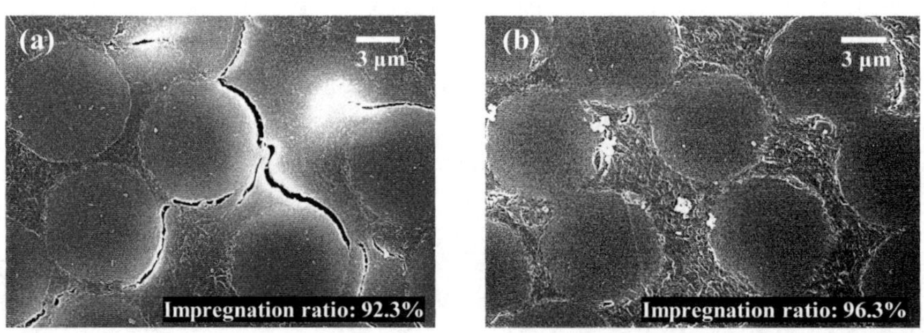


**Fig.4** SEM image of cross section of CFRTP (a) without particle, (b) with PA12 particles

*Some figures and text in this paper have been edited to make it more readable*

# EVALUATION OF EDGES FOR CARBON MATERIALS

Kazuki Matsumura, Taro Kinumoto, Tomoki Tsumuta, Masahiro Toyoda*

Department of Applied Chemistry, Faculty of Engineering, Oita University,
700 Dannoharu, Oita 870-1124, Japan.
*toyoda22@oita-u.ac.jp

## INTRODUCTION

Various oxygen-containing functional groups are bonded to the edge of carbon materials. These functional groups influence the surface chemical properties of the carbon materials; accordingly, functionalization of the carbon materials has been studied for the various applications of the carbon materials. However, there are few reports regarding differences in functionalization of edges of various carbon materials. Therefore, in this study, the oxidation treatment was applied to various materials, and the trend of changes in the number of functional groups and edges was investigated. Samples were oxidized at different times using the same treatment, and the number of edges and oxygen-containing functional groups were measured and compared via the temperature-programmed-desorption (TPD) and temperature-programmed-oxidation (TPO).

## MATERIALS AND METHODS

Exfoliated Carbon fibers (ExCFs) prepared from polyacrylonitrile (PAN)-based carbon fibers and activated carbon (AC) through desorption of functional groups by holding at 1100 °C under inert atmosphere were used as samples. In addition, ExCFs and AC were oxidized at 80 °C for 20 min or 2 h in a mixed solution of hydrogen peroxide and sulfuric acid. The sample codes are expressed adding the oxidation temperature time of the starting material such as ExCFs-0, ExCFs-20min, ExCFs-20s, AC-0, AC-20min and AC-2h. The edges were quantitatively evaluated through TPD and TPO. For the TPO measurement, a Karl Fischer moisture meter was used to measure the amount of water that was generated from sample by heating it to 1000 °C under a mixed atmosphere of He and $O_2$ through dewatering process. In addition, TPD measurements were conducted to quantify the volume of CO and $CO_2$ released from carbon samples during heating under He atmosphere. The amount of each functional group was calculated from the volume of CO and $CO_2$, and the peak area ratio was determined by the waveform separation of the TPD curve.

## RESULTS AND DISCUSSION

**Table 1** shows the volume of $H_2O$ released from each sample, as detected by TPO. It was observed that 734.2 mol $g^{-1}$ and 307.7 mol $g^{-1}$ of $H_2O$ was released from ExCFs-0 and AC-0 respectively. Thus, it was revealed that hydrogen species were present on the sample edges. Furthermore, an increase in hydrogen species was confirmed in case of each sample through oxidation by an acid solution, which is considered to correspond to an increase in oxygen-containing functional groups containing hydrogen.

Table 1 Amount of $H_2O$ released from each sample by TPO.

| Sample | ExCFs-0 | ExCFs-20min | ExCFs-2h | AC-0 | AC-20min | AC-2h |
|---|---|---|---|---|---|---|
| $H_2O$ / mmol $g^{-1}$ | 734.2 | 1115.1 | 1368.9 | 307.7 | 679.1 | 1215.8 |

**Table 2** shows the amount of CO and $CO_2$ released from each sample via TPD. In addition, **Figs. 1 and 2** show the $CO_2$ release behavior of each sample. Both ExCFs-0 and AC-0 had a trace number of functional groups, which increased after the oxidation treatment by an acid solution. However,

Table 2 Amount of CO and CO2 released from each sample by TPD.

| Sample | ExCFs-0 | ExCFs-20min | ExCFs-2h | AC-0 | AC-20min | AC-2h |
|---|---|---|---|---|---|---|
| CO / μmol $g^{-1}$ | 2.7 | 63.6 | 58.1 | 5.0 | 96.9 | 45.1 |
| $CO_2$ / μmol $g^{-1}$ | 17.1 | 22.1 | 293.8 | 22.8 | 376.0 | 355.6 |

***Some figures and text in this paper have been edited to make it more readable***

the trend of increase was different. Although ExCFs demonstrated no significant increase in the number of functional groups after oxidation treatment for 20 min, the number of functional groups derived from $CO_2$ mainly increased after oxidation for 2 h. On the contrary, in case of AC-0, the number of functional groups derived from $CO_2$ increased mainly due to the 20-min oxidation treatment; moreover, no further increase due to the 2 h oxidation treatment was confirmed. Therefore, it was demonstrated that the number of functional groups for ExCFs-0 increased for 2 h oxidation treatment with no change for a further increase in time. **Table 3** shows the number of edges for each sample. The number of edges was estimated as 1.47 mmol $g^{-1}$ for ExCFs-0; in the case of ExCFs-20min, the number of edges increased without significant increase in the number of functional groups. In addition, the number of edges increased to 2.77 mmol $g^{-1}$ for 2h oxidation treatment. Conversely, in the case of AC-0, the number of the edges increased from 0.62 mmol $g^{-1}$ to 2.43 mmol $g^{-1}$ for the 2 h oxidation treatment. In future, we will compare and examine the changes in the number of edges and functional groups for another carbon sample using the oxidation treatment for a time period greater than that tested herein.

Table 3 Amount of edges of each sample.

| Sample | ExCFs-0 | ExCFs-20min | ExCFs-2h | AC-0 | AC-20min | AC-2h |
|---|---|---|---|---|---|---|
| Edges / mmol $g^{-1}$ | 1.47 | 2.24 | 2.77 | 0.62 | 1.36 | 2.43 |

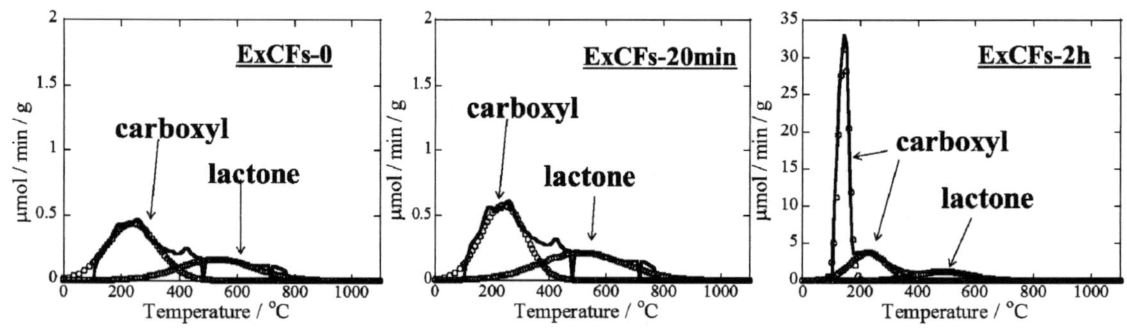


Fig.1 TPD curve corresponding to CO2 desorption for ExCFs, ExCFs-20min and ExCFs-2h.

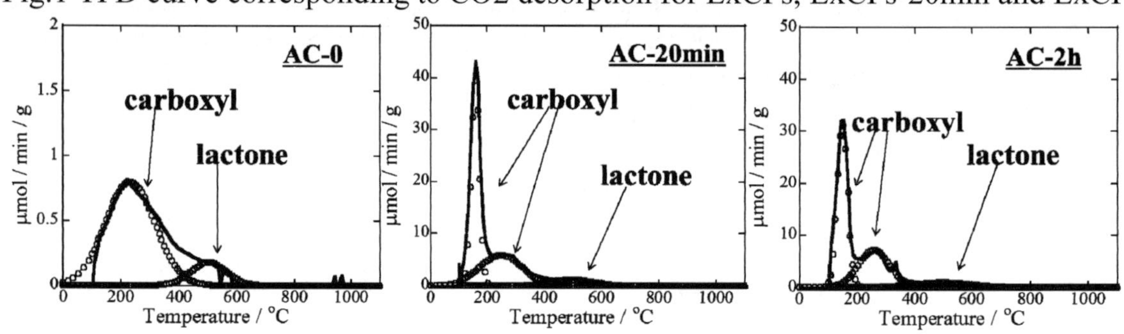


Fig.2 TPD curve corresponding to CO2 desorption for AC, AC-20min and AC-2h.

## CONCLUSIONS

The edges and the oxygen-containing functional groups of oxidation-treated samples were determined. The results indicated that the edges of ExCFs and AC increased in proportion to oxidation time. Moreover, the number of functional groups of ExCFs increased significantly after oxidation for 2 h. In contrast, the functional groups of AC increased significantly after the 20 min oxidation treatment. Thus, it was confirmed that increase in the edges does not necessarily correspond to increase in the functional groups; moreover, the oxidation time required for increasing the functional group varies depending on the measurement sample.

# Fabrication and anti-ablation performance of W/Cu composite coatings by SAPS

Ling-jun Guo*, Jian Peng
The State Key Laboratory of Solidification Processing, C/C Composites Technology Research Center, Northwestern Polytechnical University, Xi'an 710072, China
*email: guolingjun@nwpu.edu.cn

## INTRODUCTION

Tungsten/copper (W/Cu) composites, which combine the high thermal property of Cu and the low expansion and high melting point characters of W. At high temperature, Cu phases has low melting point and high boiling point, it can be used for removing heat from the engine components by a mechanism like sweating, which is beneficial to improving the anti-ablation property. Thus, applying W/Cu coating opens up a new way to increase the anti-ablation ability of C/C composites [1].

## MATERIALS AND METHODS

### Material

Specimens used as substrates were cut from a bulk 3D-C/C composites with a density of 1.75 g/cm$^3$. Tungsten powder (purity higher than 99.8%) and copper powder (purity higher than 99.5%) were purchased as feedstocks.

### Coating fabrication and Ablation Tests

SiC inner layer was prepared on the surface of C/C composites beforehand by pack cementation. Tungsten powder and copper powder were mixed and then agglomerated using a spray dryer to ensure the flow ability of particles. A supersonic plasma spraying system was used to prepare the W/Cu composite coating. An oxyacetylene ablative test with a heat flux of 2.4 MW/m$^2$ and 4.2 MW/m$^2$ were carried out to evaluate the ablation resistance property of the W/Cu composite coatings according to GJB323A-96.

## RESULTS AND DISCUSSION

Fig. 1a and Fig. 1c shows the linear and mass ablation rates of W/Cu coated composites after ablation. Considering of the requirement of dimensional stability to the application, W12Cu coating shows the best anti-ablation property, its linear ablation rate decreased by 63% under heat flux of 2.4 MW/m$^2$, 87% under heat flux of 4.2 MW/m$^2$ compared with pure W coating. The surface temperature of W/Cu coated samples as a function of ablation time are shown in Fig. 1b and Fig. 1d. It can be found that the highest surface temperature of W/Cu coated composites was lower than pure W coated composites, continuously sweating and cooling effect of Cu phase make the surface temperature of W/Cu coating decrease by 300□ at most.

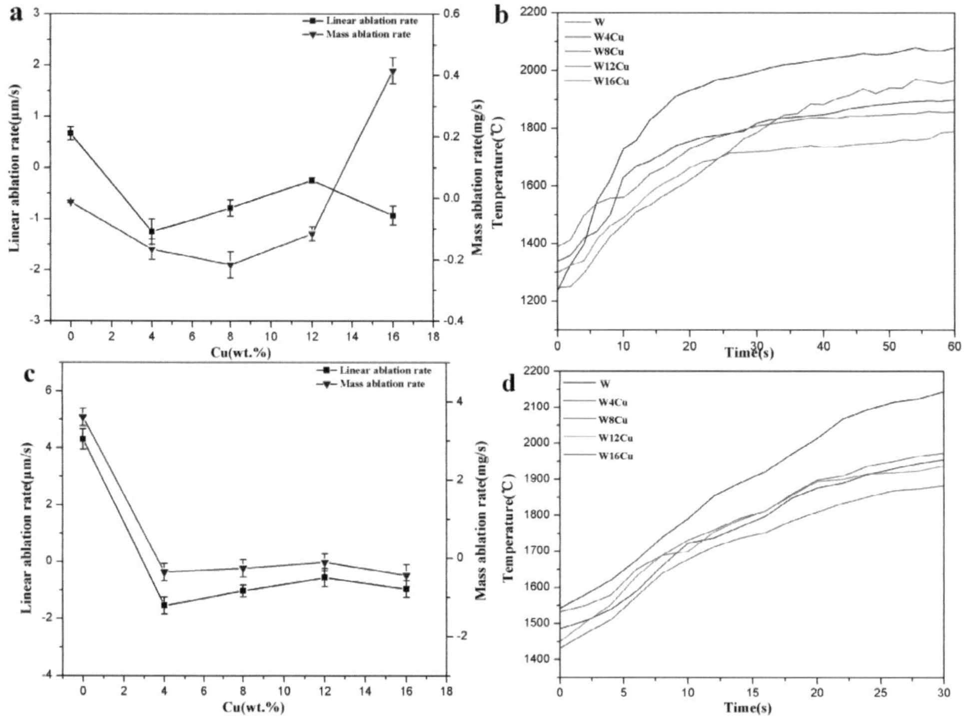


Fig. 1. ablation rates of W/Cu coated composites after ablation and surface temperature (a) and (b) under heat flux of 2.4 $MW/m^2$, (c) and (d) under heat flux of 4.2 $MW/m^2$.

Fig. 2 shows the surface SEM images of W12Cu coating in ablation center region. Due to the molten Cu film that formed on the surface of the coating during ablation, which resisted mechanical and oxidation erosion of oxyacetylene torch, W12Cu coating was dense and smooth after ablation.

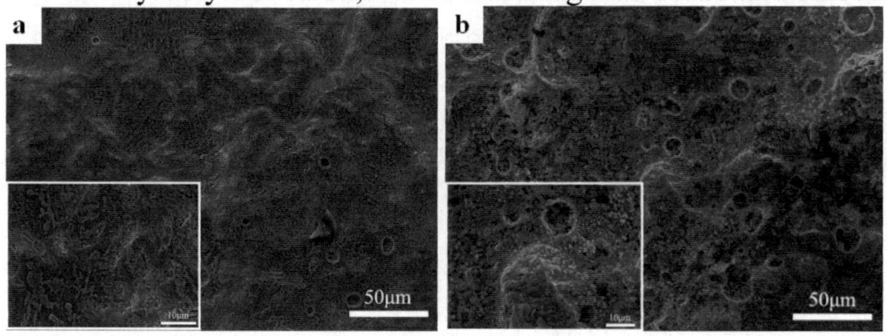


Fig. 2. Surface SEM images of W12Cu coating in the center region after ablation (a) under heat flux of 2.4 $MW/m^2$ for 60 s, (b) under heat flux of 4.2 $MW/m^2$ for 30 s.

## CONCLUSIONS

During the ablation process, molten Cu film was formed on the surface of W/Cu coating, continuously sweating and cooling effect of Cu phase decreased temperature evidently, which lead to excellent anti-ablation property of W/Cu coating.

## REFERENCES

[1] G. Pintsuk, I. Smid, J.E. Doring, et al., Fabrication and characterization of vacuum plasma sprayed W/Cu-composites for extreme thermal conditions, J. Mater. Sci. 42 (2007) 30-39.

# Fe-N co-doped graphene as air-cathode catalyst in microbial fuel cells

Dingling Wang[1], Zhaokun Ma[1]*, Yang'en Xie[1], Huaihe Song[1]

[1]*State Key Laboratory of Chemical Resource Engineering, Beijing Key Laboratory of Electrochemical Process and Technology for Materials, Beijing University of Chemical Technology, Beijing 100029, China*
**mazk@mail.buct.edu.cn*

## INTRODUCTION

A microbial fuel cell (MFC) is a promising technology to convert organic compounds to electricity using bacteria[1][2]. There are so many reports with different reference states about the application of various graphene in MFCs, having made it hard to compare and interpret their catalytic performance among these systems. Thus, we use Fe-N doped single-layer graphene, multilayer graphene and GO as cathode catalysts, and compared the electrochemical properties of the catalysts under the same condition, providing solution for optimal selection of cathode catalyst in MFCs. In the present study, we provided a simple and efficient method for the formation of CNTs on the SLG sheets. This study may throw light on the modification of new catalyst's structure to improve the electrogenesis capacity in MFCs by Fe-N catalysis,

## MATERIALS AND METHODS

### Material

a water solution containing 20 mM $FeCl_3 \cdot 6H_2O$ and 20 mM melamine[2]; graphene powders; Carbon cloth (HCP330, Heseng Co.,Ltd., China) was used as the cathode; PTFE; iso-propanol; and Nafion.

### Methodes

Carbonization under nitrogen gas; liner sweep voltammetry; scanning electron microscope (SEM).

## RESULTS AND DISCUSSION

Fig.1 exhibited that the maximum power density was 1210±20 $mW \cdot m^{-2}$ (Fe-N/SLG) Obviously, Fe-N/SLG-MFC had a better electrogenesis capacity than Pt/C-MFC.

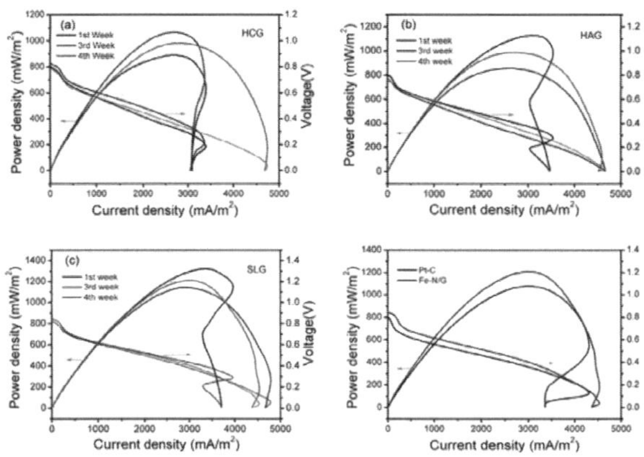


Fig. 1. Multicycle power density and polarization curves of MFCs equipped with HCG, HAG and SLG.

Some serried coralloid carbon nanotubes (CNTs) were formed in the catalyst and their diameters were estimated to be in the range of 0.1-1μm.

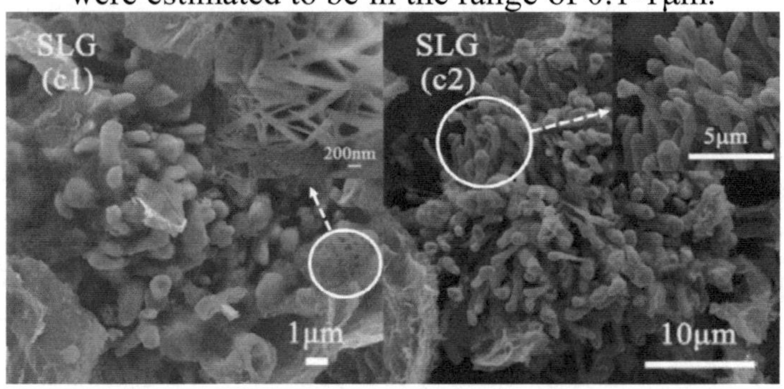


Fig. 2. SEM images of catalysts synthesized with SLG.

## CONCLUSIONS

As-prepared Fe-N/SLG catalyst with CNTs displayed superior electrogenesis capacity. It has a good application prospects in MFCs.

## REFERENCES

1. Shaffer G, Olsen S M, Pedersen J O P. "Long-term ocean oxygen depletion in response to carbon dioxide emissions from fossil fuels" *J. Nat. Geosci.* Vol.2, p.105-109.
1. Logan B E, Hamelers B, Rozendal R, et al. "Microbial fuel cells: methodology and technology" *J.Environmental Science & Technology,* Vol.40(17), p.5181-5192.
2. Jang J K, Pham T H, Chang I S, et al. "Construction and operation of a novel mediator- and membrane-less microbial fuel cell" *J. Process Biochemistry.* Vol.39(8), p. 1007-1012.

**FLEXIBLE FREE-STANDING GRAPHENE /HOLLOW TIN DIOXIDE PAPER FOR LITHIUM-ION BATTERIES**

**Wanyuan Zhi[1], Tao Xu[1], Qinghan Meng[1*], Bing Cao[1], Yunhua Yu[2]**

[1]*(College of Materials Science and Engineering, Beijing University of Chemical Technology, Beijing 100029, China)*

[2]*(State Key Laboratory of Organic-inorganic Composites, Beijing University of Chemical Technology, Beijing 100029, China)*

*(corresponding author: qhmeng@mail.buct.edu.cn)*

## INTRODUCTION

With the development of mobile electronic devices by deformation function, the deformable and flexible lithium-ion batteries for their power supply have been paid more and more attention in recent years. This article introduce a novel forming method to prepare free-standing graphene/ tin dioxide doping paper anode for lithium-ion batteries by a simply physically method, followed by freeze-drying and pressing. Graphene has a very high electrical conductivity and flexibility; it can be used as self-standing lithium-ion battery anode material all by itself. Hollow structure tin dioxide creates a rapid lithium transportation path and stable structure during charging and discharging. The structure was characterized by SEM, XRD, and the electrochemical performance was characterized by constant current charge and discharge, CV, EIS. It delivers a high discharge capacity of 930.5mAh·g$^{-1}$ at a current density of 100 mAh·g$^{-1}$ first charge, even after 100 cycles there's 548.2mAh·g$^{-1}$ when used as binder-free lithium ion battery anode. This self-standing paper with no binder or current collector has a certain mechanical strength that can be bended as an angle. This work provided the possibility for future flexible high performance energy storage devices.

## METHODOLOGY

Submillimeter sized graphene oxide solution (GOs) was synthesized by two steps as electrochemical intercalation and the modified Hummer's method[1]. The $SnO_2$ hollow microspheres were prepared by template method. 60mmol of glucose was dissolved in 60 mL deionized water, then 2mmol of SDBS was added to the solution, stirred and transferred to a 100mL Teflon-lined stainless steel autoclaver and reacted at 180°C in a blast oven for 6h then cooled to room temperature. After sonicated for 10min, 50mmol of $SnCl_2 \cdot 2H_2O$ was added to the reaction solution. After stirred well for 30min the Teflon-lined stainless steel autoclaver was heated at 180°C continuously for 1h. The resulting suspension was washed by deionized water and sieved several times, and then dried to obtain $SnO_2$ coated carbon balls precursor structure. The obtained cladding material was calcined at 450°C for 3h in a muffle furnace to remove the central carbon template to obtain hollow structure $SnO_2$ microspheres. Weighted 10mg hollow-$SnO_2$ and 5mg deionized water, then sonicated for 1h. Then added 10g GOs (5mg·g$^{-1}$) to the suspension and stirred for more than 12h to make the materials mixed evenly. The mixed suspension was transferred to 60mm flat-bottomed petri dish and frozen for 6h. After freeze-drying for 24h we obtained GO/hollow-$SnO_2$ gas gel. The air gel was pressed at a pressure of 10MPa for 2min to obtain a GO/hollow-$SnO_2$ paper, and then carbonized at 500°C in $N_2$ atmosphere for 2h to obtain the rGO/hollow-$SnO_2$ paper, signed as HSN-10. Changing the amount of hollow-$SnO_2$ to 20mg and 30mg, we got the self-standing papers signed as HSN-20 and HSN-30.

The working electrodes were fabricated easily by cut the self-standing paper to wafer whose diameter was 12mm. The as-prepared self-standing electrodes are assembled as half-cells (single electrode configuration) using CR2025 coin-cells in an argon-filled glove box for measuring the performance of lithium ion battery anode electrodes, in which lithium acts as both the counter and reference electrode, Celgard 2400 microspore membrane as the separator, and 1M $LiPF_6$ in a 1:1:1 volume ratio of ethylene carbonate/diethyl carbonate/dimethyl carbonate (EC/DEC/DMC) organic solvents as the electrolyte.

The galvanostatic charge/discharge profiles were obtained using the LAND CT2001A battery tester at various current densities. The CV and EIS were tested on Zennium electrochemical workstation.

## RESULTS AND DISCUSSION

Figure 1a shows the SEM of hollow-$SnO_2$ microspheres. The diameter of the microspheres is around 500nm and the thickness of the spherical shell is 50nm. Spherical surface is rough, which is due to the accumulation of $SnO_2$ crystalline state. From figure 1b and 1c, it can be seen that the $SnO_2$ microspheres are well dispersed uniformly in the graphene sheets and encapsulated by graphene. The cross-section SEM photo shows that the graphene and $SnO_2$ microspheres doping materials has a porous sandwich structure, and the self-standing flexible electrode has a thickness of about 100μm.This loose and porous structure increases the specific surface area of the active electrode material, which is beneficial to the immersion and diffusion of electrolyte and lithium ions in lithium-storage application. This structure also provides enough space to release the stress of $SnO_2$ volume changing in the process of charge and discharge, which protects the electrode structure from damage and improves the life of lithium battery.[2] Figure 2 shows the XRD spectrum of HSN-30. It can be seen that the diffraction spectrum of the material is in accordance with the standard pattern of $SnO_2$ (JCPDS NO.41-1455). The baseline of the spectrum appears a broad peak at 2θ=24°, which corresponds to the (002) of the carbonized GO. The spectrum shows the presence of $SnO_2$ and rGO in the material.

From figure 2a, it can be seen that the first discharge capacity of SHN-10, SHN-20 and SHN-30 is 855.6mAh·$g^{-1}$, 1167.2mAh·$g^{-1}$ and 1355.9mAh·$g^{-1}$ respectively, and the first charge specific capacity is 498.2mAh·$g^{-1}$, 696.1mAh·$g^{-1}$ and 930.5mAh·$g^{-1}$. The specific capacity and the first charge-discharge efficiency of the electrodes are improving with the increase of $SnO_2$ content. If continue to increase the doping amount of $SnO_2$, the mixture will not from a self-standing paper without binder because of the high content of powder material. HSN-30 is the best available self-standing electrode by this preparation. In the magnification performance test, HSN-30 shows the capacities of 804.3mAh·$g^{-1}$, 621.6mAh·$g^{-1}$, 382.2mAh·$g^{-1}$, 219.9mAh·$g^{-1}$ and 110.5mAh·$g^{-1}$ at 0.1A·$g^{-1}$, 0.2A·$g^{-1}$, 0.5A·$g^{-1}$, 1A·$g^{-1}$, 2A·$g^{-1}$ respectively. In figure 3c, there are slight platforms near 0.7V and 1.4V of the charging curve, which are unique charging platforms for $SnO_2$. Figure 4d shows the comparison of rGO paper electrode, hollow-$SnO_2$ and HSN-30 cycling performances. HSN-30 shows both better cycle stability than pure hollow-$SnO_2$ electrode and higher specific capacity than pure rGO paper electrode. From figure 4a, the CV curves of HSN-30, it can be seen that the oxidation peak of $SnO_2$ and $Li^+$ decompose into Sn and $Li_2O$ at 0.75V and 2.1V, and the reduction peak of the alloying process of Li-Sn at 0.7V and 1.2V. The SEI curve in figure 4b shows HSN-30 has excellent electrical conductivity and ion diffusion ability.

## ACKNOWLEDGMENT

This study is supported by National Nature Science Foundation of China (51372011).

## REFERENCES

1. Xu T, et al. (2015). RSC Advances, 5(90): 73737-73741.
2. Liang J, et al. (2012). ACS applied materials & interfaces, 4(11): 5742-5748.

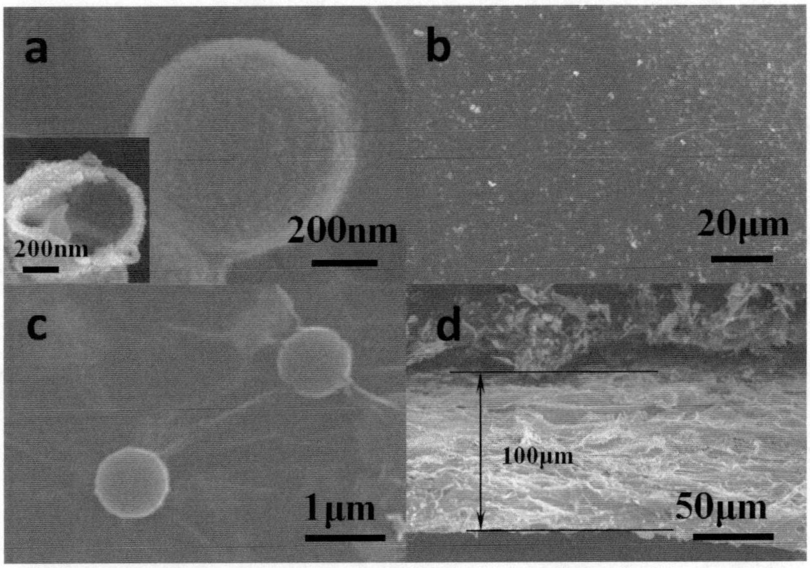


**Fig.1** The SEM images of hollow-$SnO_2$ microspheres and HSN-30. a) hollow-$SnO_2$ microspheres, b-c) the surface of HSN-30, d) the cross-section of HSN-30

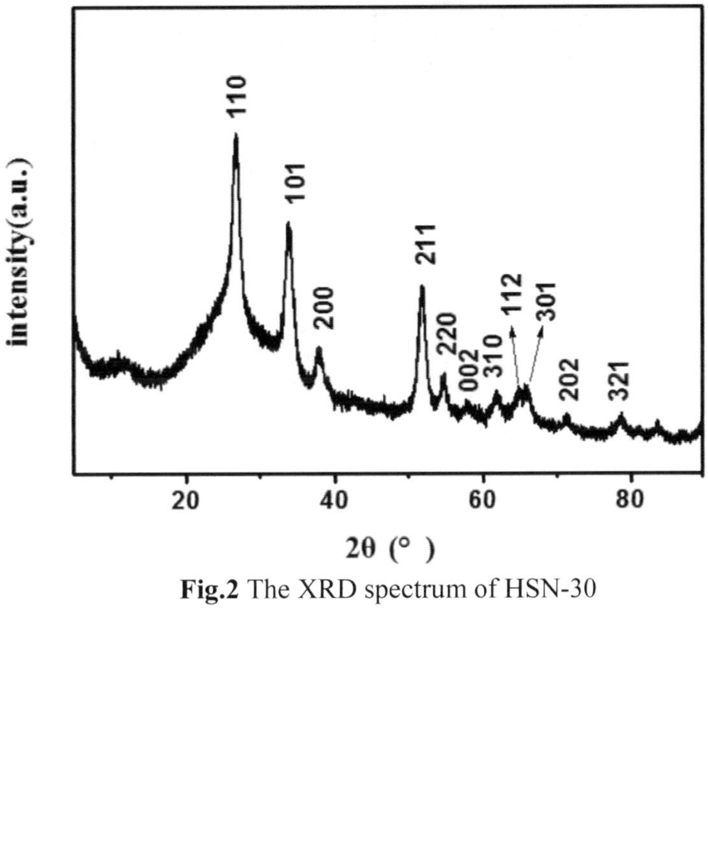


**Fig.2** The XRD spectrum of HSN-30

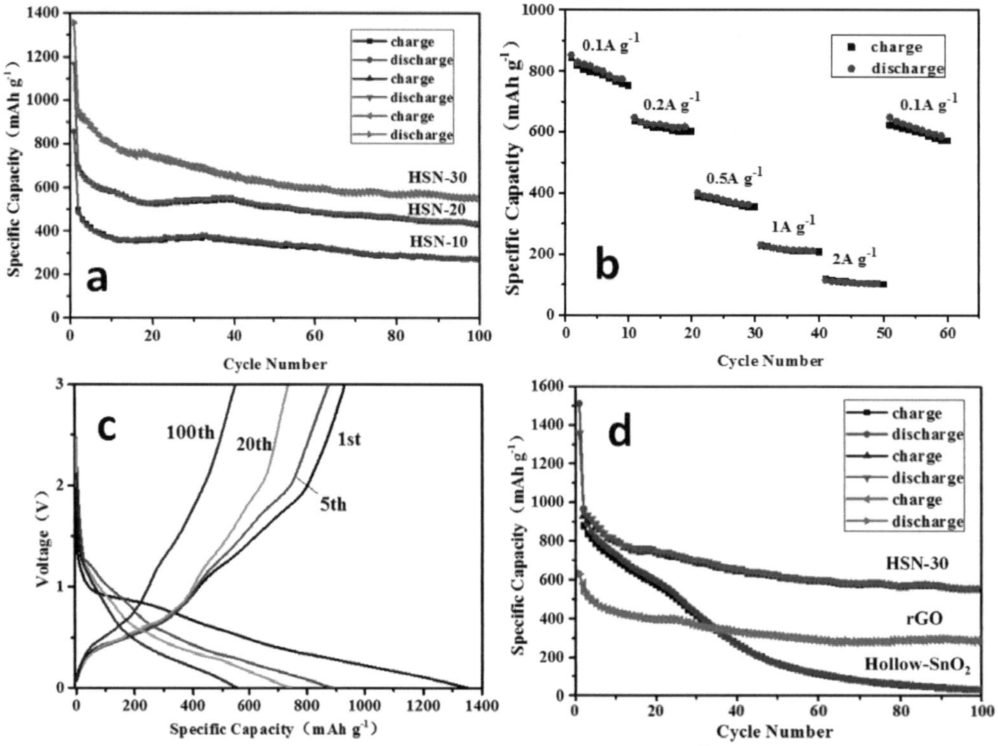


**Fig.3** a) The cycling performance of HSNs at a current density of $0.1A\cdot g^{-1}$; b) the specific versus current densities of HSN-30, c) the specific versus current densities of HSN-30; d) the cycling performance of HSN-30, rGO self-standing paper, hollow-$SnO_2$ at a current density of $0.1A\cdot g^{-1}$

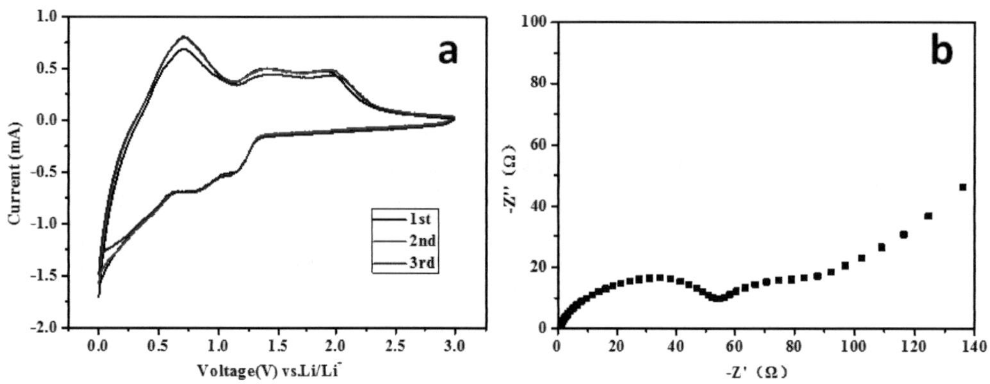


**Fig.4** a) The CV curves of HSN-30 composite anode at a scan rate of 0.2mV/s in the voltage range of 0.001-3V b) the EIS curve of HSN-30

***Some figures and text in this paper have been edited to make it more readable***

# Formation and catalytic activity for oxygen reduction reaction of fullerene-soot-derived onion-like carbons

Machiko Takigami*, Takuya Maie, Kumi Nariduka, Koji Takasu, Naokatsu Kannari, Takafumi Ishii, Jun-ichi Ozaki

Department of Science and Technology, Gunma University, Kiryu, Gunma 376-8515, Japan
*mtakigami@gunma-u.ac.jp

## INTRODUCTION

Onion-like carbons (OLCs) can be produced by several methods; intense electron irradiation of carbon soot, conversion of nanodiamonds to OLCs by heat treatment above 1600 °C, arc discharge of graphite, combustion of naphthalene and so on. OLCs are used as lubricants to reduce friction in liquid and solid states, coating for plastic films, and additives to cosmetics for UV-cut.

Fullerene soot (FS) is a carbon material for fullerene production. In this study, we report the formation of OLCs from FS by oxidative heat treatment and their catalytic activities for oxygen reduction reaction (ORR).

## MATERIALS AND METHODS

### Material

After removing fullerene from the FS produced by combustion method, the remained amorphous carbon was named and sold as Nanom black (NB, Frontier Carbon Ltd.). NB was used as a starting material for this research.

### Heat treatment

NB as received was heated in oxidative atmosphere ($O_2/N_2$=6/94) at 600 □. Weight of NB and time of heat treatment were changed.

### Characterization

Before and after heat treatment, samples were characterized by XRD, TEM, XPS and electrochemical measurements in acidic medium.

## RESULTS AND DISCUSSION

NB heated in oxidative condition decreased the weights with heating time as shown in **Fig. 1**. XRD profiles of the heat-treated samples changed as shown in **Fig. 2**. Peak around $2\theta$=14 ° disappeared and peak around 24 ° appeared and shifted to higher diffraction angles with the increase of heating time. TEMs of the samples changed as shown in **Fig. 3**. OLCs are hardly seen in pristine NB, however, OLCs which seem to be embedded in amorphous carbons can be recognized clearly with

***Some figures and text in this paper have been edited to make it more readable***

the increase of heating time. In NB heat-treated for 3 h, OLCs have several carbon layers and their diameters are less than 5 nm. However, OLCs thus prepared are not concentric circles; rather disordered and warped graphitic layers they have. Their surfaces are similar to the OLCs prepared by heat-treatment of nanodiamonds, although the OLCs prepared from nanodiamonds have more aligned carbon layers than the OLCs from NB. As the heat-treated nanodiamonds have catalytic activities for ORR[1], ORR activities of the NB heat-treated in oxidative atmosphere were measured and are shown in **Fig. 4**. ORR activities improved by oxidative treatment of NB. As the weight of NB decreases by the oxidative heat treatment, amorphous carbons in which OLCs are embedded are removed predominantly and OLCs appear. OLCs thus appeared have higher ORR activities than amorphous carbons. As NB does not contain any nitrogen, ORR activities of heat-treated NB in oxidative atmosphere can be attributed to the OLCs with disordered warped carbon layers. The ORR voltammograms of the carbons from NB had higher current densities than those of carbons nanodiamonds, indicating that the disordered warped carbon layers on the surface of carbons are responsible for the ORR if the carbons do not contain nitrogen.

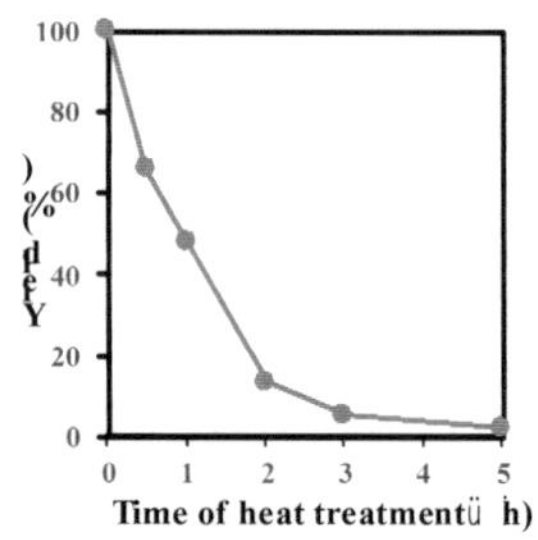


Fig. 1 Yields of NB heat-treated in oxidative

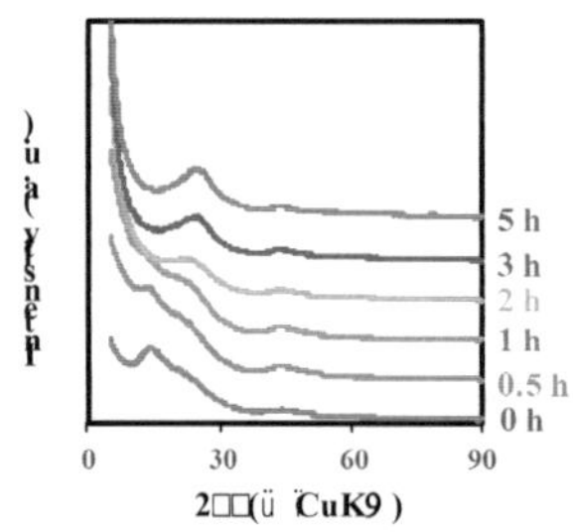


Fig. 2 XRD profiles of NB heat-treated in oxidative

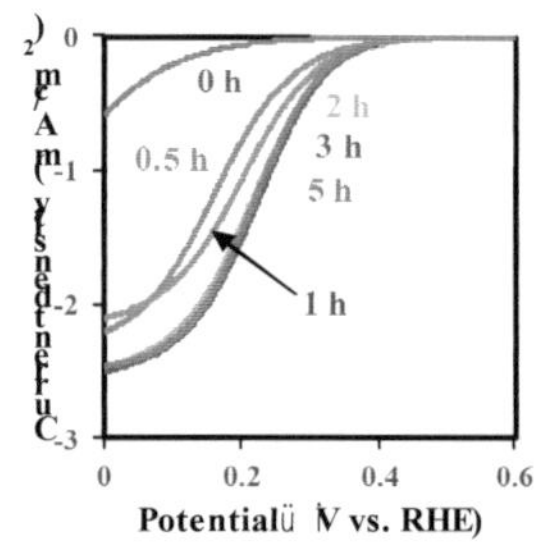


Fig. 4 ORR voltammograms of NB heat-treated in oxidative

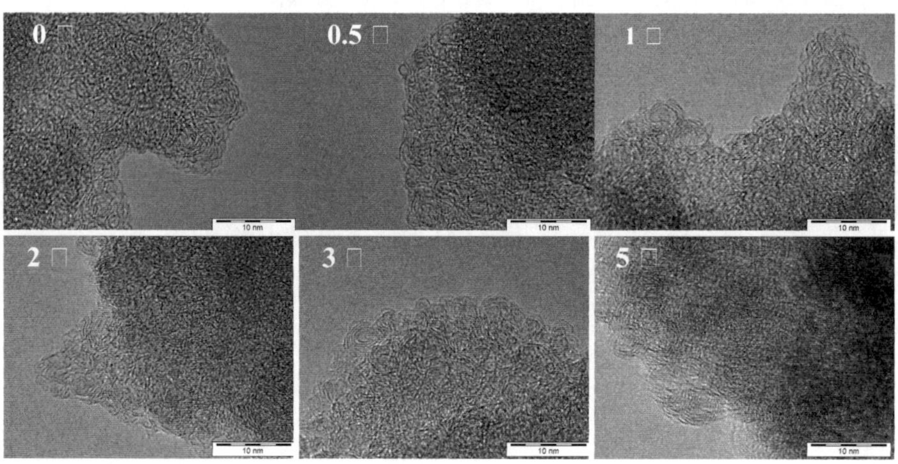


Fig. 3 TEMs of NB heat treated in oxidative atmosphere at 600 °C.

## CONCLUSIONS

OLCs were prepared by oxidative heat treatment of NB. OLCs embedded in amorphous carbons in NB appeared by removing amorphous carbons by oxidative heat treatment. The OLCs thus prepared had disordered and warped carbon layers on the surface of the carbons, and catalytic activity for ORR. Such carbon structures function as active sites for ORR.

## REFERENCES


1. N. Kannari, T. Itakura and J. Ozaki, (2015). "Electrochemical oxygen reduction activity of intermediate onion-like carbon produced by the thermal transformation of nanodiamonds", *Carbon*, Vol.87, p.415-417.

# Graphene doped based electrocatalyst for fuel cell applications

Lemes, G.[1], Luque-Centeno, J. M.[1], D. Sebastián[1], Martínez-Huerta, M.V.[2], Pastor, E.[3], Lázaro, M.J.[1*]

[1]Instituto de Carboquímica. CSIC, Zaragoza, Spain.
[2]Instituto de Catálisis y Petroquímica. CSIC, Madrid, Spain.
[3]Instituto de Materiales y Nanotecnología, Universidad de La Laguna, La Laguna, Spain.
* mlazaro@icb.csic.es

## INTRODUCTION

Recently, graphene oxide (GO) and reduced graphene oxide (rGO) based materials have been proposed to replace noble metals (as Pt) in alkaline membrane fuel cells (AMFCs) because of their high activity toward oxygen reduction reaction (ORR). Moreover, it is possible to model graphene properties through doping strategies with heteroatoms (N, P, S, B) which can modify the electronic structure. However, rGO doping procedures allow the introduction of a limited amount of heteroatoms.

On the other hand, carbon nitride (CN) has a similar structure to nitrogen doped reduced graphene oxide (NrGO) but a greater percentage of nitrogen, the thermal decomposition of CN occurs at 700 °C[1], and appears as a promising electrocatalyst[2]. Therefore, NrGO-CN composites seem to be the route to get a graphene based structure with nitrogen content as high as 35 wt%.

In this work, graphene doped based electrocatalysts with high nitrogen content have been synthesized by one step annealing process. The electroactivity towards the oxygen reduction reaction has been assessed in a three-electrode cell in alkaline medium.

## MATERIALS AND METHODS

GO was synthesized according to modified Hummer's method by exfoliation of graphite powder <20 μm, provided by Sigma-Aldrich[3]. Nitrogen doped samples were prepared via thermal annealing of a mixture of GO and urea under nitrogen atmosphere at different temperatures (500 °C, 600 °C, 700 °C and 800 °C) for 2 hours.

## RESULTS AND DISCUSSION

Elemental analysis and X-ray photoelectron spectroscopy (XPS) have been used to measure the surface chemical composition of all samples (Table 1). The nitrogen concentration diminishes with treatment temperature. The deconvolution of the XPS N1s signal (Table 1) reveals the successful incorporation of nitrogen mainly in the form of pyridinic (40-50 %) and pyrrolic (26-40%) moieties.

| Sample | C | O | N | Pyridinic N | Pyrrolic N | Graphitic N | N-oxides |
|---|---|---|---|---|---|---|---|
| GO | 49.0% | 45.9% | - | - | - | - | - |
| NrGO-CN 500 | 57.0% | 5.9% | 35.8 % | 43.6 % | 27.9 % | 25.3 % | 3.2 % |
| NrGO-CN 600 | 58.9 % | 5.6 % | 30.6 % | 40.5 % | 39.9 % | 19.5 % | 0.0 % |
| NrGO700 | 73.5 % | 2.7 % | 19.8 % | 49.7 % | 31.9 % | 16.8 % | 1.6 % |

| | | | | | | | |
|---|---|---|---|---|---|---|---|
| NrGO800 | 83.2 % | 2.4 % | 10.0 % | 44.3 % | 26.8 % | 24.7 % | 4.2 % |

**Table 1: Chemical composition in mass percentage (elemental analysis) and nitrogen speciation calculated from deconvoluted XPS N1s signal.**

The crystalline structure of the materials has been studied by X-ray diffraction (XRD). Figure 1a presents that GO has a broad peak centred at 10.8°, while the rest of samples show a main peak that is centred at ~26.5°. According to these results, all thermally treated samples have recovered the interplanar distance of graphite (26.7°).

Thermogravimetric analyses (TGAs, not shown) exhibit a significant weight loss at 650ºC of NrGO-CN. This behaviour is related to the decomposition of CN and demonstrates its presence in these composites. This sharp weight fall does not happen on NrGO700 and NrGO800.

The electrocatalytic activity was examined in a conventional three-electrode system, using a rotating disk electrode (RDE) as working electrode and oxygen-saturated 0.1 M NaOH as the electrolyte. Figure 1b compares the ORR activities of the GO-based samples and commercial Pt/C (ETEK). Linear sweep voltammetries (LSVs) show that the higher the thermal treatment temperature, the greater the ORR current density, as well as more positive onset potential.

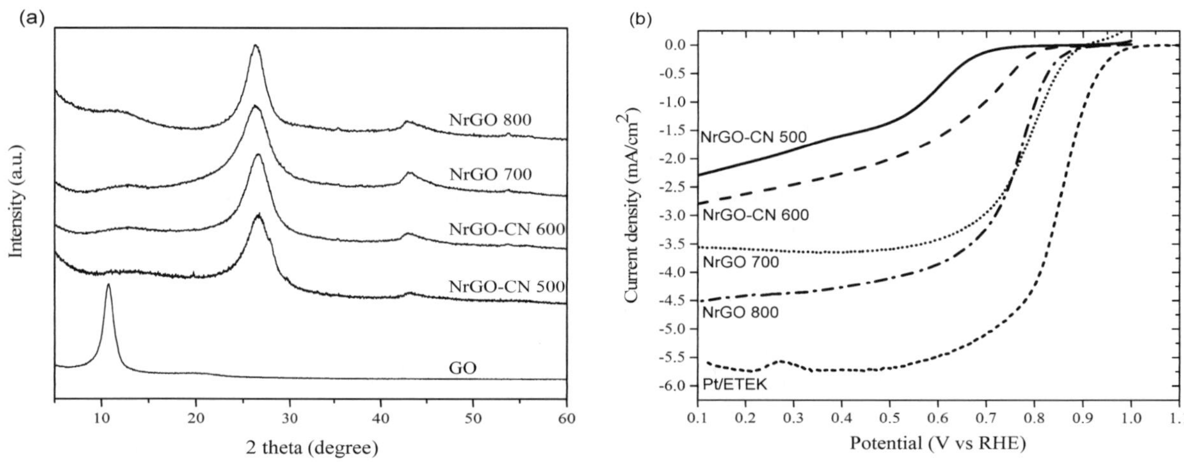


**Fig. 1. (a) XRD patterns. (b) LSVs recorded on RDE electrode in $O_2$-saturated 0.1M NaOH solution at 1600rpm. Scan rate: 2 mV/s.**

## CONCLUSIONS

NrGO and NrGO-CN electrocatalysts have been successfully synthesized and investigated by modifying the temperature of the annealing process. XPS analyses confirm the incorporation of nitrogen to the graphene structure. The electrochemical study reveals that the material treated at 800°C presents the best activity to ORR.

## ACKNOWLEDGMENT

The authors gratefully acknowledge financial support given by Spanish Ministry of Economy and Competitiveness (MINECO) through Projects ENE2014-52158-C2-1-R and 2-R. G. Lemes and J. M. Luque Centeno acknowledge Aragón Government and MINECO (FPI program), for their Ph.D. grants, respectively.

**REFERENCES**


1. Thomas, A. et al. (2008). "Graphitic carbon nitride materials: variation of structure and morphology and their use as metal-free catalysts". J. Mater. Chem., Vol 18, p.4893-4908.
2. Zhang, Y. et al. (2013). "Manageable N-doped graphene for high performance oxygen reduction reaction", Sci. Rep., Vol 3, p.2771.
3. Hummers, W. S. & Offeman, R. E. (1958). "Preparation of Graphitic Oxide". J. Am. Chem. Soc., Vol 80(6), 1339–1339.

# Graphitization of Spin Coating Polymer

Yasushi Soneda*, Noriko Yoshizawa, Masaya Kodama

National Institute of Advanced Industrial Science and Technology (AIST),
Tsukuba, Ibaraki, Japan
*y.soneda@aist.go.jp

## INTRODUCTION

High quality graphite films have commercialized from Japanese companies and used as a heat dissipation material for the potable digital devices due to its flexibility, light weight and high thermal conductivity. To synthesize the graphite films, only limited type of polymer precursors are able to be employed since they must be thermosetting and graphitizable properties. A kind of polyimide, Kapton, is known to suitable one for this purpose [1], but the precursor film needs to be prepared by casting of polyamic acid solution because of the insolubility of polyimide in solvent and results the less processability.

We have examined the preparation of polymer thin films by spin-coating method and found that poly-benzimidazobenzophenanthroline (BBL) /acid solution and polybenzoxazine (PBO) /organic solution were able to produce the superior precursor films for graphitization [2-4]. From these films prepared by spin coating on substrates, high crystallinity graphite thin films with the thickness between 50 nm to 5 μm were successfully obtained by high temperature treatment.

In the case of PBO, the high crystallinity graphite thin films were obtained when the spin coated polymer was carbonized around 600 °C with the glass substrate. This condition indicates that the stretching of spin coated film by the adhesion to the glass substrate during the carbonization period enable to enhance the graphitizability of precursor polymer. In this study, we tried the carbonization of conventional non-graphitizable precursors, such as polyacrylonitrile (PAN), furfuryl alcohol (FA), etc., by using the spin coating technique and the carbonization with stretching on the glass substrate.

## MATERIALS AND METHODS

Reagent grade of PAN (Sigma-Aldrich) was used as received. PAN (3.5 g) was dissolved with 50 ml of N,N-dimethylformamide (DMF). 400 μl of PAN solution was spin-coated (1000-4000 rpm for 3sec) on a glass substrate (2 inch) and heat treated in air at 300 °C for 4 hr (ramping rate of

1°C/min). The oxidized PAN film on the substrate was heat treated at 600 °C in nitrogen flow. Then, the carbonized film was peeled off from a substrate and graphitized in between artificial graphite plates. In the case of FA, the liquid itself was heated at 80 °C to increase viscosity and spin-coated on a glass substrate. Polystyrene/ethyl acetate solution was also tried in the similar manner with the case of PAN.

The graphite thin films thus obtained were examined with SEM (channeling contrast), XRD with reflex and transmission geometry, Raman scattering spectroscopy.

**RESULTS AND DISCUSSION**

**Fig.1** shows the photographs of oxidized PAN coating on a glass substrate, the carbonized coating at 600 °C and the graphitized at 2800 °C after the peeled off from the substrate. A clear coating on the glass substrate was obtained from PAN/DMF solution by spin coating and turned to brown colour after the air oxidation at 300 °C (**Fig. 1a**). This oxidized polymer coating was carbonized with the glass substrate at 600 °C (**Fig. 1b**). The carbonized film was able to be peeled off from substrate by immersing into water. Although the film eliminated from glass substrate was sandwiched with artificial graphite plate and graphitized at 2800 °C, the film shape was largely broken into several pieces (**Fig. 1c**).

**Fig. 1 Photographs of spin coated PAN films with different heat treatments: (a) oxidized in air at 300 °C with glass substrate, (b) carbonized film of (a) in nitrogen 600 °C and (c) graphitized in between graphite plate at 2800 °C (no substrate).**

X-ray diffraction pattern of a piece of graphitized PAN specimen on a silicon non-reflecting holder shows only 00*l* diffraction lines (**Fig. 2**). At the transmission geometry, these diffractions were not observed but only 100 and 110 diffractions were detected.

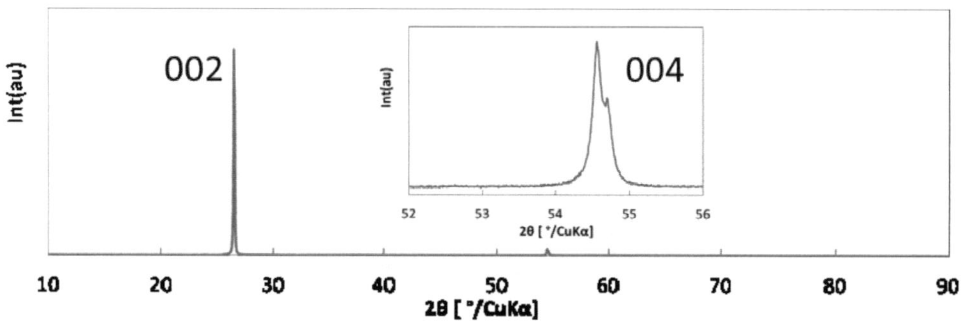


**Fig. 2 X-ray diffraction pattern of the graphitized PAN from spin coated film at 2800 °C (corresponding to a piece of Fig. 1c).**

Using the same procedure, PS and FA could produce a graphitized film even though small portion. These results indicated that the conventional non-graphitizable precursors possess the ability to produce graphite when the precursor films were carbonized with stretching.

**REFERENCES**

1. Inagaki M, et al. "High quality graphite films prepared from aromatic polyimides". In: Thrower PA, Radovic LR editor. Chemistry and Physics of Carbon, vol 26, New York; Marcel Dekker; 1999 p. 245−333.
2. Yamashita J, et al. (2010). TANSO 2010;245: 196-199 [in Japanese].
3. Soneda Y, et al, (2013). Extended Abstracts of Carbon2013, Brazil, 2013, ID362.
4. Soneda Y, et al, (2015). Extended Abstracts of Carbon2015, Dresden, 2015,PCP21.
5. Soneda Y, et al, (2015). Extended Abstracts of Carbon2016, Penn State, 2016,O4-3.

# HIGH CAPACITANCE AT HIGH SCAN RATE BY HYBRID GRAPHENE-CARBON XEROGEL SUPERCAPACITORS

G. Ramos-Fernández[1*], M. Canal-Rodriguez[2], A. Arenillas[2], J.A Menéndez[2], I.Martin-Gullón [1]
[1]Department of Chemical Engineering, University of Alicante, Spain
[2]INCAR-CSIC, Oviedo, Spain.
*gloria.ramos@ua.es

## INTRODUCTION

Electrochemical energy storage devices, such as supercapacitors (SCs), batteries and fuel cells have attracted widespread attention among the scientific community over the last few decades due to the increasing use of renewable energy and the rapid growth in demand for electricity by electronic goods and electric vehicles. Electrical double layer capacitors (EDLC), show an excellent cyclability, low maintenance and a longer life span and a safer operation record than pseudo-capacitive SC. Electrode materials are one of the most important factors for the improvement of the electrochemical performance of EDLCs, to get a higher energy storage capacity and power densities. EDLCs are based on high surface area carbon based materials, mainly porous carbons [1]. Carbon gel materials, like carbon aerogel, cryogel and xerogel based on resorcinol (R) and formaldehyde (F), offer a greater potential [2-4] than the traditional activated carbons. However, they also show a limited energy capacity and power density when the current density is increased like other porous carbons. Power density is inversely proportional to the equivalent series resistance (ESR), which is related to the electrical conductivity although its quantitative influence with respect to other factors like the BET surface area is not yet clear. In view of these problems, carbon gels must necessarily be activated in an oxidative atmosphere to maximize the BET surface area to attain the required energy densities. Furthermore, hybrid EDLCs based on carbon xerogel composites with carbon nanomaterials, like carbon nanotubes and graphene. To promote a good dispersion and avoid aggregation, graphene oxide (GO) is an excellent choice for the preparation of composite carbon materials.

## MATERIALS AND METHODS

Initially an organic gel was synthesized by the polycondensation of resorcinol with formaldehyde in water as a solvent and with sodium carbonate as gelation catalyst, based on the method described by Pekala et al.[5]. The present work carries out the synthesis of the organic gel by substituting an aqueous suspension of GO at different concentrations (10, 15, 20 mg/mL) by the water. Thus, the gel is formed using a GO sheet as a scaffold platform.

A GO-free organic gel was also synthesized for comparison purposes. These xerogels were further carbonized (CX) in nitrogen atmosphere at 800ºC attaining medium pore developments (700-800 $m^2/g$ BET), where graphene oxide is *in situ* reduced increasing the electrical conductivity. Subsequently, portions of these hybrid graphene-carbon xerogels were steam activated at two burn-offs to increase microporosity (ACX) for the improvement of the electrochemical properties, combining by the double effect of conductivity plus porosity.
Analysis of the electrochemical behavior of the materials was carried out using 1M $H_2SO_4$ as electrolyte, and applying cyclic voltammetry (CVs), in a voltage window of between 0 and 1.0 V at different scan rates, and constant charge-discharge (CD) current, with a potential fixed at 1V. The electrical conductivity of the electrodes was determined by the van der Pauw method.

## RESULTS AND DISCUSSION

Figure 1 A shows the Ragone plot, calculated from the CD at 1 A/g, for all the samples, with different amounts of GO and without any GO (with different symbols for each), and non-activated and activated samples (solid and clear, respectively). The best performance, in terms of high energy and power density, corresponds very clearly to graphene-doped non-activated xerogels, especially those with 1.2% and 1.9% in the parent organic gels. These samples, with less than 700 $m^2/g$, nearly double the performance of the equivalent non-activated pristine xerogel in terms of both energy and power density. The large difference in power density is due to big differences in ESR which are much lower in the carbonized graphene-doped xerogels than in the pristine ones. With respect to the performance of graphene-doped activated xerogels, it is close to, but slightly lower than that of the graphene doped non-activated xerogels, with an intermediate ESR. Consequently, ESR seems to play more important role than the BET specific surface area. This parameter is known to depend greatly on the electrical conductivity of the materials.

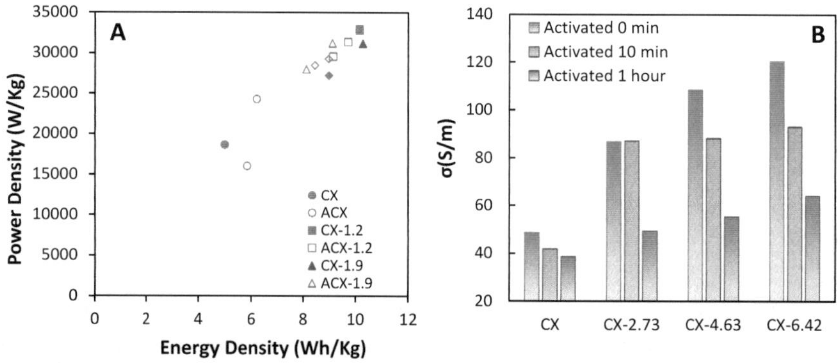


**Fig.1. A. Ragone plot for the supercapacitors, B Conductivity of non-activated and activated carbon xerogels with different GO contents.**

Figure 1 B shows the electrical conductivity for the different carbon xerogels with different contents of GO and activation degrees. Now, there is a clear tendency towards increasing electric conductivity with increasing GO content, especially for non-activated samples, where the conductivity reaches twice the value of the graphene free samples. The increase in electrical conductivity is the factor responsible for the higher values of power density in the GO containing xerogels, as shown in Figure 1 A.

## CONCLUSIONS

The main conclusion from the present work is that graphene oxide produced a significant enhancement of the electrical conductivity of the carbon xerogel resulting in a marked improvement of both the energy and, especially, power densities, with even better results when carbon xerogels were non-activated and presented only 600 $m^2/g$.

## REFERENCES

[1] Wei L, Yushin G. Nanostructured activated carbons from natural precursors for electrical double layer capacitors. Nano Energy. 2012;1(4):552-565.

[2] Calvo EG, Ania CO, Zubizarreta L, Menéndez JA, Arenillas A. Exploring new routes in the synthesis of carbon xerogels for their application in electric double-layer capacitors. Energy and Fuels. 2010;24(6):3334-3339.

[3] Calvo EG, Lufrano F, Staiti P, Brigandì A, Arenillas A, Menéndez JA. Optimizing the electrochemical performance of aqueous symmetric supercapacitors based on an activated carbon xerogel. Journal of Power Sources. 2013;241:776-782.

[4] Tsuchiya T, Mori T, Iwamura S, Ogino I, Mukai SR. Binderfree synthesis of high-surface-area carbon electrodes via CO 2 activation of resorcinol-formaldehyde carbon xerogel disks: Analysis of activation process. Carbon. 2014;76:240-249.

[5] Pekala RW. Organic aerogels from the polycondensation of resorcinol with formaldehyde. Journal of Materials Science. 1989;24(9):3221-3227.

# High-strength and Modulus Carbon Nanotube Fibers

OK-Kyung Park[1], Hoikil Choi[1], Jaesang Yu[1], Jae Kwan Lee[2], Chong Rae Park[3] Bon-Cheol Ku[1*]

[1]Institute of Advanced Composite Materials, Korea Institute of Science and Technology (KIST), Jeonbuk, Korea
[2]Department of Carbon Materials, Chosun University, Gwangju, South Korea.
[3]Department of Materials Science and Engineering, Seoul National University, Seoul, South Korea
*cnt@kist.re.kr

## INTRODUCTION

Among the carbon nanotube (CNTs)-based structural materials, carbon nanotube fibers (CNTFs) are very attractive materials in the high-performance composite industry due to their high mechanical properties, electrical conductivity, and low density [1]. In despite of the great properties of CNTs, a limitation in enhancing mechanical properties of CNTFs exists in the form of weak interfacial interactions between adjacent CNTs in the CNTFs. One promising approach to overcome this limitation is the introduction of cross-linking bonds between adjacent carbon shells [2]. It is possible to prevent the slip behaviour occurring between the CNT bundle and induce efficient load transfer when the cross-linking between the CNTs is formed, thereby dramatically improving the mechanical properties. In this study, we propose an effective method to improve mechanical properties of CNTFs by introducing cross-linking bonds between individual CNTs composing CNT fiber using low molecular weight aryl diazonium salt cross-linking agent.

## MATERIALS AND METHODS

*Fabrication of Carbon nanotube fibers (CNTFs).* The CNTFs were fabricated according to a previous report [3].

*Preparation of chemically cross-linked carbon nanotube fiber (CL-CNTFs).*0.6 g of p-diaminophenyl and 0.2 g of $NaNO_2$ were immersed in 50 mL of $H_2SO_4$ at 60 °C using a syringe. The mixture of the suspension was stirred at 60 °C for 1 h. The CNTFs were dipped into a diazonium salt solution and immersed at 60 °C. The immersion time of raw CNTFs was increased from 1 to 72 h (1, 6, 12, 24, 48, 72 h). After finishing the cross-linking reaction, the cross-linked CNTFs (CL-CNTFs) were rinsed in DI-water and dried at 60 °C for 24 h [4].

## RESULTS AND DISCUSSION

Raman analyses were utilized for confirming the cross-linking reaction of CNTFs (**Table1**). The $I_D/I_G$ ratio of raw CNTFs and CL-CNTFs was investigated with increasing reaction time. After 72h of reaction time, the $I_D/I_G$ ratio of the CL-CNTFs increased from 0.09 to 0.37 because the disorder on the surface of the CNTs increased. This result confirms the occurrence of a cross-linking reaction.

An FAVIMAT single fiber tester was used to investigate the effects of the aryl cross-linking groups on the mechanical properties of the CL-CNTFs. Consequently, when the CNTFs were cross-linked for sufficient time over 24 h, significant enhancements in the tensile strength and Young's modulus were observed; the tensile strength increased from 1.6 to 3.7 N/tex, and Young's modulus increased from 46 to 210 N/tex, i.e., enhancements of mechanical properties approximately 130 % and 350 %, respectively, compared to the pristine CNTFs (see **Table 1**). In contrast, CNTFs that reacted for 24 h without diaminobenzene exhibited no improvement in mechanical properties compared with the pristine CNTFs. These results demonstrate that aromatic linker between CNTFs can significantly enhance mechanical properties. Because the mechanical properties of CNTFs greatly depend on the load transfer between the individual CNTs, such superior mechanical properties can be attributed to the inter-tube cross-linking bonding between the individual CNTs. The inter-tube cross-linking between the individual CNTs may induce efficient load transfers between adjacent CNTs and also it can prevent the slip behaviour [5], thus improving the mechanical properties of the CNTFs.

**Table 1.** $I_D/I_G$ ratio, mechanical properties and Tex of commercial carbon fiber, raw CNTF and CNTF-$H_2SO_4$ and CL-CNTF [4].

| Sample | $I_D/I_G$ ratio | Tensile strength [N/tex] | Young's modulus [N/tex] | Elongation [%] | Tex |
|---|---|---|---|---|---|
| Raw CNTF | 0.09 | 1.6 ± 0.41 | 46.1 ± 10.3 | 6.6 ± 0.9 | 0.07 ± 0.012 |
| CNTF-$H_2SO_4$ | 0.12 | 1.8 ± 0.56 | 49.8 ± 13.6 | 4.4 ± 0.5 | 0.07 ± 0.015 |
| CL-CNTF-1h | 0.19 | 2.1 ± 0.27 | 97 ± 8.5 | 3.1 ± 0.5 | 0.10 ± 0.011 |
| CL-CNTF-6h | 0.20 | 2.4 ± 0.16 | 110 ± 14.4 | 3.2 ± 0.4 | 0.11 ± 0.012 |
| CL-CNTF-24h | 0.31 | 3.4 ± 0.58 | 201 ± 7.5 | 3.7 ± 0.5 | 0.08 ± 0.015 |
| CL-CNTF-72h | 0.37 | 3.7 ± 0.42 | 210 ± 8.0 | 3.0 ± 0.6 | 0.08 ± 0.008 |

## CONCLUSIONS

We presented an effective novel method to make high-strength and modulus CNTFs via aryl radical cross-linking reactions. The results showed that the cross-linking reaction of CNTFs can improve the mechanical properties of resulting CNTFs due to the interconnection of adjacent CNTs in the CNTFs. Compared to the pristine CNTFs, the tensile strength increased to 3.7 N/tex, and the modulus increased to 210 N/tex, i.e. enhancements of mechanical properties of approximately 130% and 350%, respectively. These results clearly confirm that the aryl cross-linking reaction is an effective method of developing mechanically strong CNTFs.

## ACKNOWLEDGEMENT

This work was funded by grants from the Korea Institute of Science and Technology (KIST) Open Research Program, and Industrial Fundamental Technology Development Program (10052838) by the Ministry of Trade, Industry and Energy (MOTIE) of Korea.

## REFERENCES

1. Choo, H., Jung, Y., Jeong, Y., Kim, H. C., Ku, B.-C.(2012). "Fabrication and applications of carbon nanotube fibers", *Carbon Letters* Vol.13, p.191-204.

2. Beese, A.M., Wei, X., Sarkar, S., Ramachandramoorthy, R., Roenbeck, M.R., Moravsky, A., Ford, M., Yavari, F., Keane, D.T., Loutfy, R.O., Nquyen, S.T., Espinosa, H.D.(2014). "Key factors limiting carbon nanotube yarn strength: exploring processing structure-property relationship." *ACS Nano*, Vol. 8, p. 11454-11466.

3. Jung. Y., Kim, T., Park, C.R.(2015). " Effect of polymer infiltration on structure and properties of carbon nanotube yarns." *Carbon*, Vol. 88, p. 60-69.

4. Park, O.-K., Choi, H., Jeong, H., Jung, Y., Yu, J., Lee, J. K., Hwang, J.Y, Kim, S. M., Jeong, Y., Park, C. R., Endo, M., Ku, B. -C.(2017). "High-modulus and strength carbon nanotube fibers using molecular cross-linking", *Carbon*, Vol.118, p.413-21.
5. Cai, J.Y., Min, J., McDonnell, J., Church, J.S., Easton, C.D., Humphries, W., Lucas, S., Woodhead, A.(2012). "An improved method for functionalization of carbon nanotube spun yarns with aryldiazonium compouds." *Carbon*, Vol.50, p.4655-4662.

# Hydrogen transfer phenomenon on carbon materials with different edge structures

Daichi Umeda, Takaaki Togo, Yasuhiro Yamada,* Satoshi Sato

Chiba University, Chiba, Japan
*y-yamada@chiba-u.jp

## INTRODUCTION

Chemisorption of hydrogen on carbon materials is expected to be applied to hydrogen storage, but the detailed mechanism of hydrogen transfer phenomenon has not been clarified[1,2]. In this study, we aimed to elucidate the hydrogen transfer phenomenon on the edge of carbon materials by analyzing hydrogenation and dehydrogenation of various aromatic compounds.

## EXPERIMENTAL

### Procedure

9,10-Dihydroanthracene (DHA), 9,10-dihydrophenanthrene (DHP) and adamantane were sealed in a glass ampoule tube under reduced pressure and heated at 573-803 K to examine dehydrogenation temperature from $sp^3$C-H to $sp^2$C-H. In addition, hydrogenation experiment of anthracene, tetracene and triphenylene was conducted by heating hydrogenated aromatic compounds at 773 K under a hydrogen pressure of 0.9 MPa. For example, tetracene hydrogenated at 0.9 MPa and 773 K was abbreviated as Tetrance 0.9_773. These samples were analyzed using elemental analysis and diffuse reflectance infrared Fourier transform spectroscopy (DRIFTS). Transition states of dehydrogenation and hydrogenation were calculated using DFT calculation (Gaussian 09, b3lyp/6-31 g (d)).

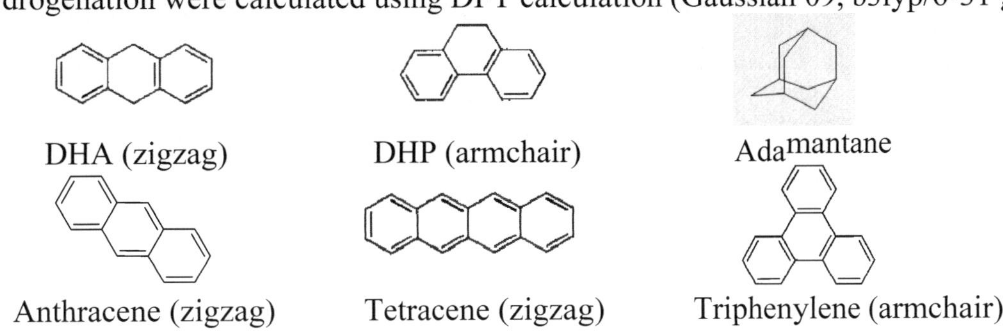


Fig. 1 Structures of compounds used in this work.

## RESULTS AND DISCUSSION

Figure 2 shows the DRIFT spectra of adamantane, DHA and DHP heated at 603-803 K. In the adamantane heated at 803 K, the peak derived from $sp^3$C-H stretching vibration at 2850-2950 $cm^{-1}$ disappeared and only the peak derived from $sp^2$C-H stretching vibration around 3050 $cm^{-1}$ appeared. At a temperature of 783 K or less, no big difference in DRIFT spectrum was observed, but discoloration occurred at 743 K or higher, and elemental analysis showed a decrease in the number of hydrogen atoms above 763 K. Disappearance of $sp^3$C-H peak was confirmed at 723 K for DHA and 773 K for DHP. Thus, it was found that $sp^3$C-H of adamantane is more difficult to dehydrogenate by heating. Since DHA dehydrogenated at the lowest temperature (723 K), it was found that zigzag type structure was the most liable to dehydrogenate.

Figure 3 shows the DRIFT spectra of anthracene, tetracene and triphenylene heated at 773 K under a hydrogen pressure of 0.9 MPa. In the tetracene, a peak derived from sp$^3$C-H newly appeared around 2800-3000 cm$^{-1}$, indicating that the hydrogenation progressed. Moreover, the shape of the DRIFT spectrum of hydrogenated tetracene matched with that of 5,12-dihydrotetracene, therefore, it is considered that the 5, 12 positions were hydrogenated first. This result was consistent with the results of DFT calculation that the activation energies of hydrogenation at positions 5 and 12 are the lowest compared to other positions and the zigzag type structure was the most suitable for hydrogen storage. Hydrogenation could not be confirmed when using triphenylene with armchair type structure or anthracene as a raw material. On the other hand, hydrogenation could be confirmed at 723 K in pentacene (not shown in Figure), suggesting that hydrogenation tends to proceed with increasing the number of aromatic rings.

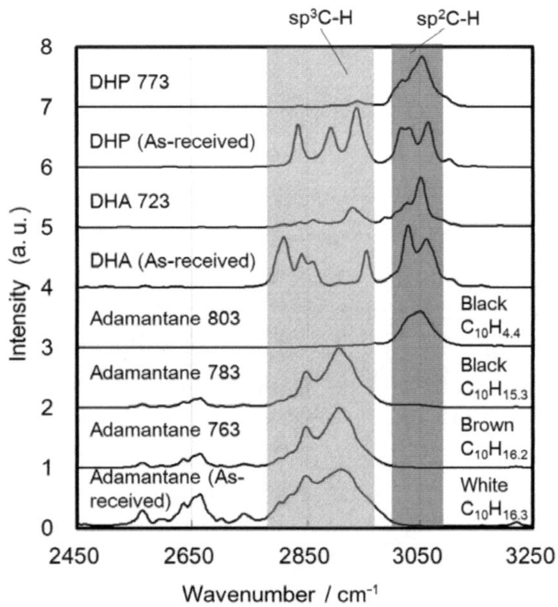


Fig. 2 Actual DRIFT spectra of heat-treated adamantane, DHA and DHP.

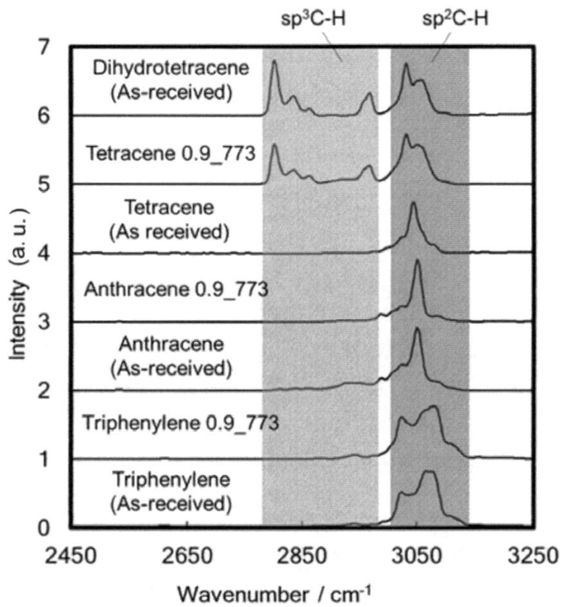


Fig. 3 Actual DRIFT spectra of PAHs, hydrogenated PAHs and dihydrotetracene.

## CONCLUSIONS

We found that the zigzag type structure tends to be dehydrogenated the most, since sp$^3$C-H of DHA was the easiest to be dehydrogenated. Even in hydrogenation, zigzag type compounds showed the high reactivity. From these results, it is concluded that the tetracene-like zigzag type structure is the most effective for chemical adsorption and desorption, and structures with a larger number of rings are desirable.

## REFERENCES


1. A. J. Lachawiec, Jr., G. Qi and R. T. Yang (2005), "Hydrogen Storage in Nanostructured Carbons by Spillover: □ Bridge-Building Enhancement", *Langmuir*, Vol.21(24) p.11418-11424

2. F. H. Yang, A. J. Lachawiec, Jr. and R. T. Yang (2006) "Adsorption of Spillover Hydrogen Atoms on Single-Wall Carbon Nanotubes", *J. Phys. Chem. B*, Vol.110(12), p.6236-6244

## Influence of out-of-plane compression induced damage effects on the mechanical properties of C/C

Behnisch, T., Thieme, M., Böhm, R., Gude, M.

Presenting author's e-mail: robert.boehm1@tu-dresden.de

*Institute of Lightweight Engineering and Polymer Technology, Technische Universität Dresden, Dresden, Germany*

Carbon fibre reinforced carbon (C/C) is a composite material with a high temperature resistance. Because of its high weight specific rigidity and strength, C/C is preferably used for complexly loaded high temperature applications [1, 2]. The inner structure of C/C on the mesoscale is characterised by stacked textile layers made of carbon fibres with a bidirectional orientation. Hence, this composite material shows good in-plane properties and the material behaviour is well predictable by a variety of material models which were developed in recent years. Due to a comparatively material intrinsic high porosity caused by emerging cracks and outgassing during the high temperature manufacturing process, C/C is compressible under local out-of-plane compression load. Such mechanical stress situations primarily occur in technical systems at positive-fitting and friction-locked connection points or local material interfaces of hybrid structures. The examined carbon fibre reinforced carbon is manufactured by Schunk Kohlenstofftechnik GmbH with fabric reinforcement (bidirectional 0°/90° fibre orientation) and an open porosity between 8 and 10 %. The inner structure with 0° and 90° layers at the mesoscale is shown in Figure 1 (left and middle) and as an inverted picture of the translaminar void system (Figure 1, right). The material is compressible up to 11 % under out-of-plane compression load.

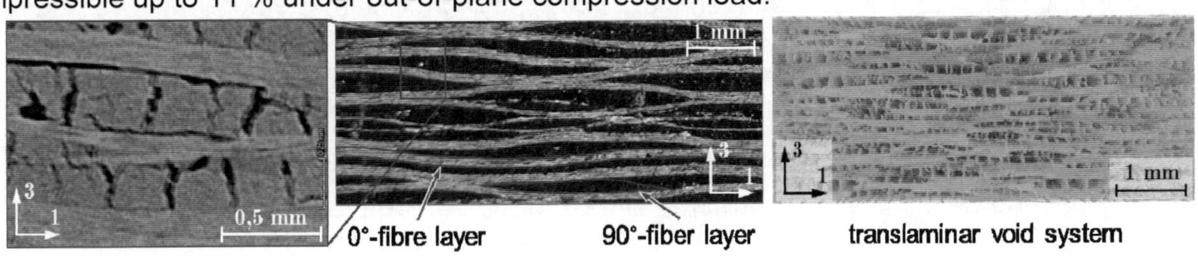


**Figure 1: Inner structure of the examined carbon fibre reinforced carbon**

With increasing deformation under local out-of-plane compression, damage effects occur, and they results in a significant degradation of the mechanical properties which leads to a changed in-plane material behaviour. Initially, an experimental in-situ investigation of the inner material micro structure was conducted while the compressive load was applied. Using x-ray computer tomography and optical micro section analysis, the emerging out-of-plane damage effects caused by inner deformations or relative movements of the composite components were identified [3, 4]. The observed damage effects are schematically shown in Figure 2. Based on the local out-of-plane compression load, inter alia an interfacial failure between the fibre layers occurs. This failure mode is caused by shear stresses as a result of differences of the lateral strain properties of the 0° and 90° fibre layer (cf. Figure 2, left). Moreover, inter-fibre fracture due to matrix failure or failure of the bonding between fibre and matrix emerge (Figure 2, middle). The reason for this failure mode is an out-of-plane compression induced shear stress in the 90° fibre layers. Additionally, fibre fracture in the 0° fibre layer occurs (Figure 2, right). This failure mode is a result of the shear effect between two separated 90° fibre layer fragments in consequence of the inter-fibre fracture in combination with an increasing local out-of-plane compression load.

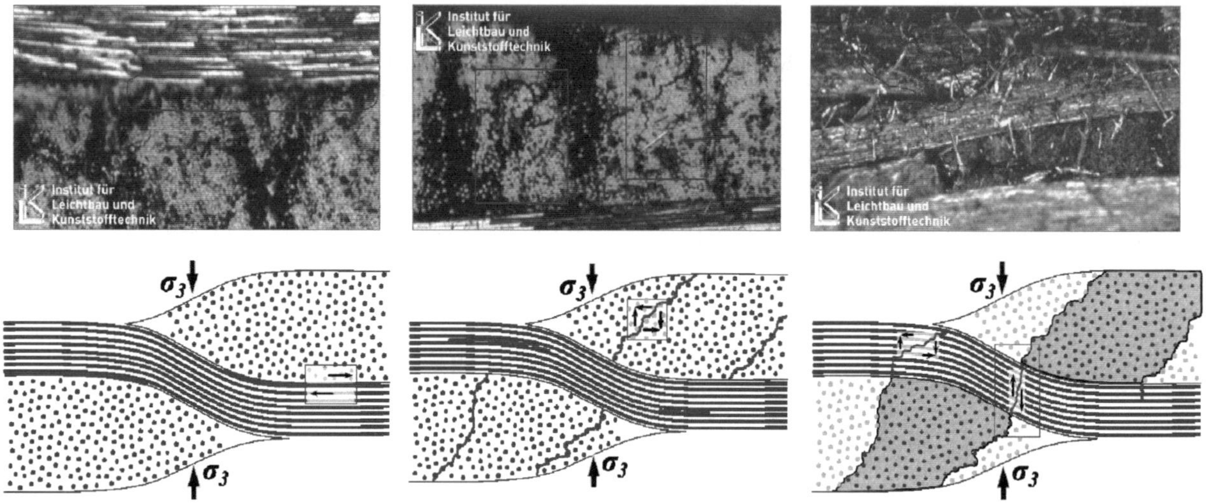


**Figure 2: Out-of-plane compression induced damage effects in carbon fibre reinforced carbon**

The influence of these out-of-plane damage effects on the in-plane behaviour of C/C is currently not addressed in material modelling yet. This leads to an overassessment of the structural strength and the load bearing capability of structures made of carbon fibre reinforced carbon. Hence, a modified material model to describe the impact of local out-of-plane compression loading on the in-plane mechanical properties is provided. This damage model describes the change of the in-plane material behaviour – the change of *damage initiation*, *damage evolution* and *failure behaviour* in particular. For the description of the damage initiation, a stress based damage approach is used [5]. To consider the influence of local out-of-plane compression, the strength for the damage initiation $R_{D,i}$ is written as a function of the local out-of-plane compressive stress $\sigma_3^{(-)}$. With the function shown in Figure 3, the strength degradation is modelled and the stress-strain-curve is controlled by the model parameters $u_{D,i}$ and $v_{D,I}$ (point 1). Relating to the damage evolution, the degradation of stiffness $C_i$ is taken into account using the damage parameter $D_i$ (Figure 3, point 2). The damage evolution is described by a damage growth function $\Phi_i$. The damage parameter $D_i$ is described by the sum of the in-plane damage growth and the damage growth induced by the out-of-plane compression load (Figure 3, point 2). With the parameter $q^i_{3(-)}$, the influence of the out-of-plane compression damage on the damage modes $i$ is taken into account. To model total failure, a failure mode based criterion from Cuntze is used [6]. The influence of out-of-plane compression on the failure strengths $R_i$ is determined similar to the damage initiation modelling by a formulation as a function of $\sigma_3^{(-)}$ (Figure 3, point 3).

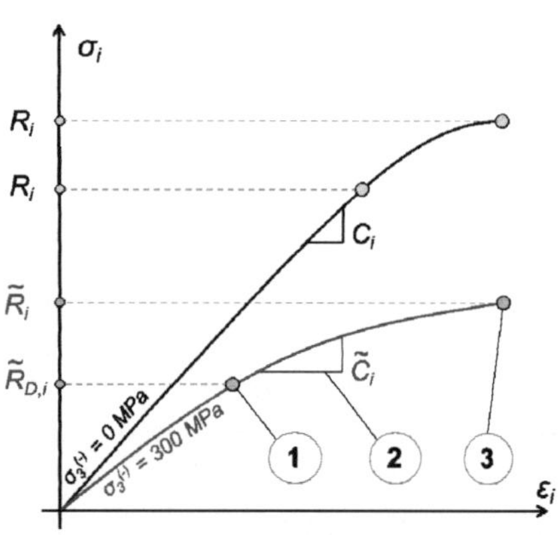


(1) damage initiation: $(i = 1, 2, 12)$

$$\tilde{R}_{D,i} = R_{D,i}\left(\sigma_3^{(-)}\right) = \left(\frac{1}{1+\left(\frac{\sigma_3^{(-)}}{\sigma_i}\right)^{u_{D,i}}}\right)^{\frac{1}{u_{D,i}}} \cdot R_{D,i} \cdot \left(\frac{\sigma_3^{(-)}}{\sigma_i}\right) \cdot \frac{R_3^{(-)} - v_{D,i} \cdot \sigma_i}{R_{D,i}}$$

(2) damage evolution: $(i = 1, 2, 12)$

$$\tilde{C}_i = (1 - D_i) \cdot C_i \qquad (D_i)^n = (\phi_i)^n + \left(\phi_3^{(-)} \cdot q^i_{3(-)}\right)^n$$

(3) failure behaviour: $(i = 1, 2, 12)$

$$\tilde{R}_i = R_i\left(\sigma_3^{(-)}\right) = \left(\frac{1}{1+\left(\frac{\sigma_3^{(-)}}{\sigma_i}\right)^{u_i}}\right)^{\frac{1}{u_i}} \cdot R_i \cdot \left(\frac{\sigma_3^{(-)}}{\sigma_i}\right) \cdot \frac{R_3^{(-)} - v_i \cdot \sigma_i}{R_i}$$

**Figure 3: Modelling approach for the out-of-plane compression induced damage effects on the in-plane material behaviour**

**Literature**


[1] Neubrand, A.: Hochtemperaturwerkstoffe mit attraktivem Eigenschaftsprofil. In: *Konstruktion - Zeitschrift für Produktentwicklung und Ingenieur-Werkstoffe* 7/8 (2014), pp. IW8 – IW10

[2] Pietig, R. ; Dürr, M. ; Lewalter, A. ; Bathe, C. ; Chrost, W. ; Michels,M. ; Henrich, M. ; Weiß, R. ; Feltin, D. ; Preussler, S. ; Ackermann, D.; Matthäus, G. ; Hufenbach, W. ; Richter, H. ; Weck, D. ; Gruner, H. ; Moens, J. ; Linke, J. ; Pintsuk, G. ; Bobzin, K. ; Schläfer, T. ; Warda, T. ; Thao, L.: Ultra fast rotating anodes for high power X-ray tubes. In: *High Temperature Ceramic Materials and Composites* AVISO, Berlin, 2010, pp. 652 –657

[3] Hufenbach, W.; Behnisch, T.; Langkamp, A.: Effect of damage by compression loading in third direction on the mechanical properties of CMC. In: *9th X-Ray Forum* Hannover, Wunstorf, 2013, pp. 20 – 22

[4] Böhm, R. ; Stiller, J. ; Behnisch, T. ; Zscheyge, M. ; Protz, R. ; Radloff, S. ; Gude, M. ; Hufenbach, W.: A quantitative comparison of the capabilities of in situ computed tomography and conventional computed tomography for damage analysis of composites. In: *Composites Science and Technology* 110 (2015), pp. 62 – 68

[5] Böhm, R.: Bruchmodebezogene Beschreibung des Degradationsverhaltens textilverstärkter Verbundwerkstoffe, Technische Universität Dresden, PhD Thesis, 2008

[6] Cuntze, R.G.: Failure conditions for isotropic materials, unidirectional composites, woven fabrics - their visualisation and links. In: *Proceedings of the Conference on Damage in Composite Materials (CDCM)* Stuttgart, 2006

# Interconnected Fe-Mn fibrils on activated carbon to increase As(V) uptake

J.J. Gutiérrez-Martínez [1*]; J.R. Rangel-Méndez [1**]

[1]División de Ciencias Ambientales, Instituto Potosino de Investigación Científica y Tecnológica A.C., San Luis Potosí, México.
[*]joel.gutierrez@ipicyt.edu.mx , [**]rene@ipicyt.edu.mx

## INTRODUCTION

The environmental pollution caused by the presence of arsenic in drinking water has driven the development of highly selective materials for As(V) adsorption[1]. To address these issues, the impregnation of iron oxyhydroxide on activated carbon (AC) by a microwave assisted hydrotermal (MH) process is known as the best alternative to enhance the effectiveness of AC by tailoring its specific properties in order to increase its affinity for As(V). An alternative to improve the uptake of AC is by means of the addition of chemical reagent like Mn (sometimes called foreign ion) through iron oxyhydroxide anchorage, which allows to modifying the crystalline morphology, influencing in the nature of the reaction products of iron [2]. The foreign ions act as structure modulators, promoting the formation of different crystal habits [3]. This effect allows increasing the quantity and availability of active sites on iron oxyhydroxide. Therefore, the challenge of this work is to further evaluate the effect of Mn on the crystalline morphology of iron oxyhydroxides anchored on the activated carbon surface, and to obtain an adsorbent material with enhanced As(V) adsorption capacity by anchoring interconnected Fe-Mn fibrils.

## MATERIALS AND METHODS

AC was modified at different concentration of $FeSO_4 \cdot 7H_2O$: $MnSO_4 \cdot H_2O$ solution through MH process. To evaluate the physical and chemical properties of the materials, these were characterized by phisisorption of $N_2$, acidic digestion, potentiometric titration, XRD and SEM. As(V) adsorption experiments were carried out in a bath system at pH 7 and 25°C.

## RESULTS AND DISCUSSION

Characterization results reveal that iron modified AC under hydrotermal condition did not severely reduce their porosity (Table 1). Interestedly, the presence of Mn during iron impregnation allows loading a largest amount of iron on the carbon surface (3.47 %) without affecting the porosity. Textural properties of Mn-F400 disclosed that effectively Mn did not precipitate on AC. The foregoing result suggests that at the acidic conditions present during MH process (pH < 3) the graphitic layer were oxidized improving the porosity. This finding also implies that acidic conditions make the surface charge of the activated carbon more negative after the MH process, which results in the shift of the $pH_{PZC}$ to lower pH values. SEM images showed an atypical behavior of the iron oxyhydroxide crystal growth (Fig. 1). To our knowledge, this structure has not been previously reported by AC modified with iron oxyhydroxide; we named it interconnected fibrils habit. The arrangement of the interconnected fibrils allowed forming pores with a diameter of 2 nm (yellow arrows), which tally with the increase in the pore volume of Fe:Mn-2M. The fibrils development is attributed to the presence of Mn, which modulates the structure of iron oxyhydroxide and promoted twinned crystals. Adsorption experiment revealed that As(V) uptake by iron-modified AC increased markedly when Mn was used as foreign ion during iron anchorage (Fig. 2), since the quantity and availability of active sites was higher in Fe:Mn-2M than in Fe-F400.

**Table 1.** Textural properties, iron content and $pH_{PZC}$ of adsorbent materials

| Sample | $S_{BET}$ ($m^2/g$) | $V_{total}$ ($m^3/g$) | $V_{ultram}$ ($m^3/g$) | $V_{micro}$ ($m^3/g$) | Iron content (%) | $pH_{PZC}$ |
|---|---|---|---|---|---|---|
| F400 | 776.4 | 0.42 | 0.12 | 0.15 | 0.41 | 9.2 |
| Mn-F400 | 796.4 | 0.43 | 0.12 | 0.15 | 0.27 | 4.6 |
| Fe-F400 | 703.3 | 0.38 | 0.10 | 0.14 | 2.51 | 4.6 |
| Fe:Mn-2M | 734.8 | 0.40 | 0.09 | 0.16 | 3.47 | 4.8 |

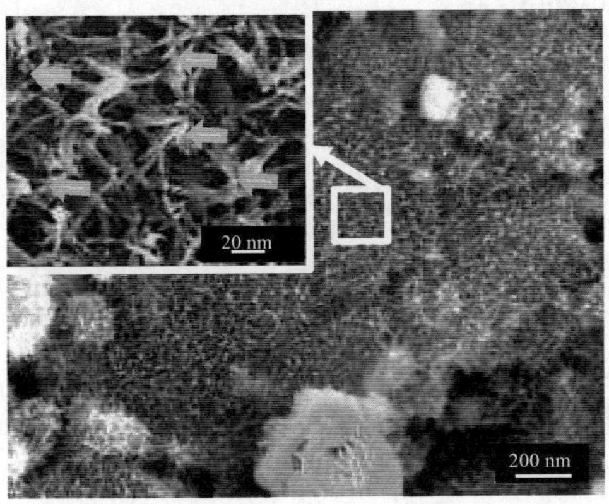


**Fig. 1.** Micrographs of sample Fe:Mn-2M.

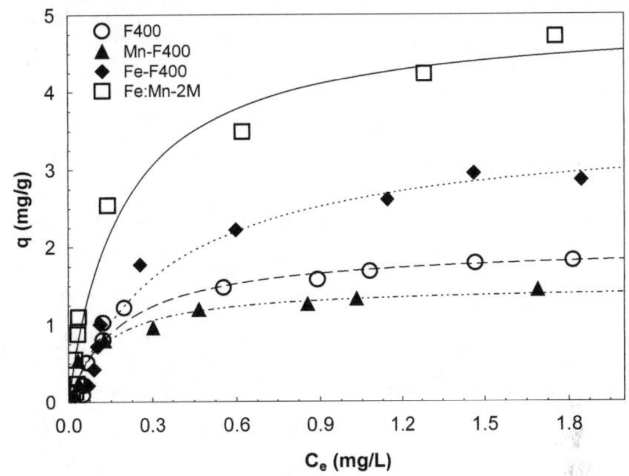


**Fig. 2.** As(V) adsorption isotherms of studied materials at 25 °C and pH 7. Lines represent the Langmuir model.

## CONCLUSIONS

The presence of Mn during the iron oxyhydroxides synthesis, under hydrothermal conditions, allowed to form crystal habits with a high concentration of available active sites that improves the As(V) uptake. The AC iron content increases from 0.41 to 3.47% after its modification, without affecting its textural properties. In Addition, microscopy studies confirmed that Mn defined the crystallization microenvironment of the iron oxyhydroxide morphology. Moreover, this element limits the crystal growth on some faces to generate crystalline habits as interconnected fibrils.

## ACKNOWLEDGE

The authors appreciate the financial support of project Ciencia Básica 2014-CONACYT-237118.

## REFERENCES

1. Mohan, D., Pittman, C.U., 2007. Arsenic removal from water/wastewater using adsorbents-A critical review. J. Hazard. Mater.
2. Jolivet J-P., Henry M., Livage J., and B.E., 2000. Metal Oxide Chemistry and Synthesis.
3. Asenath-Smith, E., Estroff, L.A., 2015. Role of Akaganeite (β-FeOOH) in the Growth of Hematite (α-Fe2O3) in an Inorganic Silica Hydrogel. Cryst. Growth Des.

# Investigation of chemical states of nitrogen introduced to fullerene-soot-derived onion-like-carbons and its property toward oxygen reduction reaction

Takafumi Ishii*, Takuya Maie, Kumi Nariduka, Naokatsu Kannari, Machiko Takigami, Jun-ichi Ozaki
Gunma University, Kiryu, Japan
*ishii@gunma-u.ac.jp

## INTRODUCTION

Nitrogen doped carbons are some of the most promising candidates for cathode catalysts in polymer electrolyte fuel cells. To be developed as high performance catalysts, it is important to identify the active sites of nitrogen doped carbons for oxygen reduction reaction (ORR). Pyridinic structures have been reported as one of the candidates for the active sites. However, chemical states of nitrogen are complicated and still be under debate because their states are affected not only by the chemical structure of nitrogen but also by many factors, such as defects and a curvature of graphene sheets. In this study, nitrogen doped onion-like carbons (N-OLCs) derived from fullerene soot were prepared as a model material for defects and/or curvature induced nitrogen doped carbons. The chemical states of nitrogen were investigated by using a high-temperature temperature-programmed-desorption (H-TPD) technique, as well as its relation toward the ORR activity was discussed.

## MATERIALS AND METHODS

*Sample preparation.* Onion-like carbon (OLC) was obtained by heat-treating Nanom Black (NB, Frontier Carbon Ltd.) at 600 °C under oxidative atmosphere (6 % $O_2/N_2$). The N-OLC was prepared by heat-treating OLC under 50 % $NH_3/N_2$ at 600 °C for 1 h.

*Characterization.* Quantitative analysis for surface functional groups on the carbon samples is performed by means of a H-TPD technique. The details of the TPD apparatus and the experimental procedure were described in our previous report [1]. Briefly, a carbon sample (~ 1 mg) was placed into a graphite holder in a quartz chamber and evacuated to high vacuum. When the pressure reached under 2 × 10-5 Pa, the carbon sample in the holder was heated by a high-frequency induction coil from room temperature to 1600 ºC at a heating rate of 10 ºC $min^{-1}$ under high vacuum. Gases (CO, $CO_2$, $N_2$, HCN, $H_2O$ and $H_2$) desorbed from the sample during a TPD run were quantitatively analyzed by a quadrupole mass spectrometer using previously-obtained calibration curves and fragment ratio data for various gases.

*Electrochemical measurement.* Cyclic voltammetry (CV) curves were obtained in nitrogen saturated 0.5 mol $dm^{-3}$ $H_2SO_4$ aqueous solution. The catalytic activity of the prepared carbons in the ORR was evaluated by linear sweep voltammetry (LSV) employing an oxygen saturated 0.5 mol $dm^{-3}$ $H_2SO_4$ aqueous solution after the CV measurement. Voltammograms were obtained by sweeping the potential from 1.0 V vs. reversible hydrogen electrode (RHE) to 0 V vs. RHE at 1 mV $s^{-1}$ by rotating the electrode at 1600 rpm. Double layer capacitance was compensated for by subtracting the voltammogram obtained in nitrogen-saturated electrolyte from the obtained ORR voltammogram.

## RESULTS AND DISCUSSION

Figure 1 shows H-TPD profiles for the carbon samples. OLC releases larger amounts of CO, $CO_2$ and $H_2O$ gases than N-OLC, which are produced by decomposing surface oxygen groups. This suggests that oxygen functional groups in OLC were removed by the nitrogen doping process. Desorption of the nitrogen species (HCN and $N_2$) are also confirmed in the profile of N-OLC. The amount of nitrogen species is larger than that of the oxygen-containing gasses. The results of H-TPD measurements suggests that the nitrogen doping induces nitrogen functional groups as well as removing the surface oxygen groups.

The results of electrochemical measurements are illustrated in Figure 2. Redox peaks are confirmed in the CV curve of OLC, resulting from pseudocapacitance provided by oxygen functional groups on the surface. The peaks decrease after the nitrogen doping. These results agree with that of the H-TPD measurements. Figure 1b shows the LSV curves obtained in the acidic solution for the carbon samples. N-OLC shows higher activity for ORR than the non-doped one, indicating that nitrogen doping induces an enhancement of the activity even when the carbon has defects and/or curvature. In our presentation, the chemical states of nitrogen formed on N-OLCs will be further investigated by using X-ray photoelectron spectroscopy, H-TPD and quantum chemical calculation.

## REFERENCES


[1] Ishii T, Kashihara S, Hoshikawa Y, Ozaki J, Kannari N, Takai K, et al. Carbon 2014;80:135-45.

# Magnetic Carbon Nanosphere-Based Surface Molecularly Imprinted Polymers for Dibenzothiophene Adsorptive Removal

Lei Qin[1,2], Weifeng Liu[1,2], Yongzhen Yang[1,2] *, Xuguang Liu[1,3] *, Lanqing Hu[4]

[1]Key Laboratory of Interface Science and Engineering in Advanced Materials, Ministry of Education, Taiyuan 030024, China
[2]Research Center of Advanced Materials Science and Technology, Taiyuan University of Technology, Taiyuan 030024, China
[3]College of Chemistry and Chemical Engineering, Taiyuan University of Technology, Taiyuan 030024, China
[4]College of Materials Science and Engineering, Taiyuan University of Technology, Taiyuan 030024, China
*liuxuguang@tyut.edu.cn; yyztyut@126.com

## INTRODUCTION

Surface molecular imprinting technique provides an efficient way to adsorptive deslufurization, which can not only effectively ease the environmental concerns but also make full advantages of the valuable organic sulfide. In this work, porous magnetic carbon nanospheres ($Fe_3O_4$@C) were obtained and employed as support materials to optimize the solid-liquid separation process after entire adsorption. After that, the surface molecularly imprinted polymers (SMIPs) were synthesized as a kind of deslufurization sorbent. Field emission scanning electron microscopy (FESEM), transmission electron microscopy (TEM) and vibrating sample magnetometry were used to characterize the morphologies and structures of the products. Furthermore, The adsorption capacity, selectivity and regenerability of SMIP/$Fe_3O_4$@C were investigated by gas chromatography.

## MATERIALS AND METHODS

### Material

Silane coupling agent KH-570 (95%), methacrylic acid (98.5%), initiating agent AIBN (≥ 99%), ethylene glycol dimethacrylate (98%), dibenzothiophene (DBT, 98%), benzothiophene (BT, ≥ 98%), deionized water, ethanol and other organic solvent were used.

### Methods

$Fe_3O_4$@C were obtained by solvothermal method. After that, the SMIPs were synthesized on the $Fe_3O_4$@C through a series of functionalization, grafting and elution processes with methacrylic acid as the functional monomer, ethylene glycol dimethacrylate as cross-linker and DBT as templete molecules to acquire a deslufurization sorbent. At last, the adsorption performances of SMIP/$Fe_3O_4$@C towards DBT in n-hexane solution (3 mmol $L^{-1}$, 298 K) were detected by gas chromatography.

## RESULTS AND DISCUSSION

The particle size of SMIP/$Fe_3O_4$@C is about 130 nm with nice dispersibility and uniformity. The saturation magnetization values of $Fe_3O_4$@C and SMIP/$Fe_3O_4$@C are 32.2 and 28.4 emu $g^{-1}$, respectively, which is in favor of regeneration after adsorbing targeted molecules under the external magnetic field.

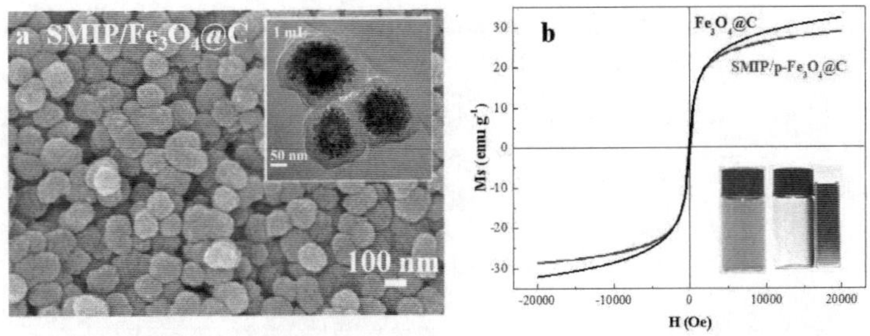


**Fig.1.** (a) FESEM image of SMIP/$Fe_3O_4$@C (inset image: TEM image of SMIP/$Fe_3O_4$@C); (b) magnetization curves of parent $Fe_3O_4$@C and SMIP/$Fe_3O_4$@C (inset image: magnetic separation of SMIP/$Fe_3O_4$@C in water, 298 K)

The saturated adsorption capacity of SMIP/$Fe_3O_4$@C towards DBT reaches 15.2 mg $g^{-1}$, much higher than that of NIP/$Fe_3O_4$@C and $Fe_3O_4$@C when equilibriums are reached in 60 min. SMIP/$Fe_3O_4$@C also exhibits better selectivity towards DBT than NIP/OMCNs does. Besides, the adsorption capacity of SMIP/$Fe_3O_4$@C declines only about 20% after 5 cycles, indicating its good regeneration ability.

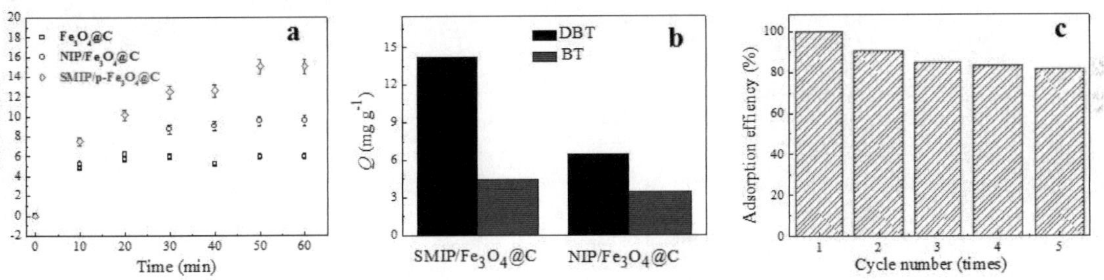


**Fig.2.** (a) Kinetic curves for adsorption of DBT dissolved in octane solution on parent $Fe_3O_4$@C, NIP/$Fe_3O_4$@C and SMIP/$Fe_3O_4$@C at 298 K;
(b) selective adsorption of DBT with BT onto SMIP/$Fe_3O_4$@C and NIP/$Fe_3O_4$@C;
(c) reusability of SMIP/$Fe_3O_4$@C

## CONCLUSIONS

The SMIP/$Fe_3O_4$@C sized around 130 nm shows adsorption capacity of 15.2 mg $g^{-1}$ and good selectivity over competitive compounds. Additionally, the adsorbent can be well recycled under magnetic field owning to the introduction of magnetic support.

## ACKNOWLEDGMENT

This work was supported by the National Natural Science Foundation of China (U1610225, U1607120, 51603142), Shanxi Provincial Key Innovative Research Team in Science and Technology (2015013002-10, 201605D131045-10), Natural Science Foundation of Shanxi Province (201601D021043), and Key Laboratory of Carbon Materials opening project, ICCCAS (KLCMKFJJ1702).

## REFERENCES

1. Liu W. F., Qin L., Shi W. P., et al. (2016). "Molecularly imprinted polymers on the surface of porous carbon microspheres for capturing dibenzothiophene", *Microchim. Acta*, Vol.183, p.1153-1160
2. Qin L., Shi W. P., Liu W. F., et al. (2016). "Surface molecularly imprinted polymers grafted on

ordered mesoporous carbon nanospheres for fuel desulfurization", *RSC Adv.*, Vol.6, p.12504-12513
3. Liu J. Z., Wang W. Z., Xie Y. F., et al. (2011). "A novel polychloromethylstyrene coated superparamagnetic surface molecularly imprinted core-shell nanoparticle for bisphenol A", *J. Mater. Chem.*, Vol.21, p.9232-9238

**Microwave assisted synthesis of graphene-$Bi_8La_{10}O_{27}$-Zeolite nanocomposite with efficient photocatalytic activity towards organic dye degradation**

**Yonrapach Areerob, Won-Chun Oh***

*Department of Advanced Materials Science & Engineering, Hanseo University, Chungnam 356-706, South Korea*

(E-mail: wc_oh@hanseo.ac.kr)

## INTRODUCTION

Organic dyes are widely used as coloring agents in industries such as paper, textile, plastic, and food. During the dyeing process, about 15% of the total world production of dyes is lost, and discharged into industrial effluents every day. Recently, considerable interest has been shown in graphene-based materials decorated with metal/metal oxide nanoparticles for the treatment of organic contaminants in wastewater through a photocatalytic technique. This treatment can completely destroy the pollutants, and has broader potential applications. Recently, the bismuth-based semiconductor photocatalyst, an important new photo-catalytic material, has attracted considerable attention, due to its remarkable photocatalytic activities for the degradation of organic compounds. In this study, we report the synthesis of G-$Bi_8La_{10}O_{27}$-Z by a microwave-assisted combustion method. This method enables a fast and uniform heating rate that causes rapid nucleation of particles and growth, which can reduce the reaction time of the irradiation. Therefore, it can consume a large amount of energy, and in turn facilitates making porous materials for photocatalytic activities. The use of this nanostructure for photodegradation and adsorption is established for the simultaneous removal of RhB, MB, and TXB dye solutions under UV light illumination [1].

## METHODOLOGY

The Hummers' Method was applied for the preparation of graphene oxide [2]. For synthesis of GBZ, 0.2 g of GO was dispersed in 100 mL of DI water. After that, for further exfoliation of graphene oxide and to form a homogenous solution, mild ultrasonication was performed for 45 min. Then, 0.136 g of $Bi_8La_{10}O_{27}$ was gradually dissolved in this GO solution to form G-$Bi_8La_{10}O_{27}$ (GB) composite. Typically, GBZ composite is fabricated by a modified sol-gel process, in which 0.1g of GB is fully dissolved in ethanol solution. Meanwhile, diluted nitric acid and water with a volume ratio of 1:5 was mixed. Subsequently, the desired amount of zeolite (2 g) was added to the above-mentioned gel Then, a yellowish xerogel was formed after aging for 2 h, and drying in a oven. Finally, GBZ composite was yielded by annealing the xerogel in a muffle.

## RESULTS AND DISCUSSION

The patterns in the XRD Spectrum of $Bi_8La_{10}O_{27}$, comprising peaks at the 2θ values of 28-50°, correspond to the (001), (002), (101), (102), (110), (112), (113), (220) and (310) planes. All the diffraction patterns of $Bi_8La_{10}O_{27}$ matched those of the JCPDS data card No. 89-4388 [3]. The surface morphology of the GBZ was investigated by SEM and TEM. Fig show micrography of the nanocomposite of GB, clearly illustrating the nano-flakes of graphene oxide and the nano-crystals of zeolites. The large surface areas of the uniformly distributed GO nanosheets can also be observed. Moreover, Fig 4 show the results of the photocatalytic performance of the prepared GB, respectively, which were evaluated from the photocatalytic decomposition of the RhB, MB, and TXB dyes. The degradation ratio (Ct/Co) of GBZ15 is much higher than that of the other catalysts used to produce dyes of RhB, MB, and TXB,

## CONCLUSION

In this study, GBZ15 photocatalyst was successfully synthesized by adopting a simple and green microwave irradiation technique in aqueous medium. GBZ15 was then utilized for the photocatalytic degradation of RhB, MB, and TXB dye molecules under natural sunlight irradiation.

## References:


*1.* M. Ghanbari, M. Bazarganipour, M. S. Niasari, *Separation and Purification Technology, Volume 173, January 2017, Page 27–36.*

[1] R. Sahraei, Z. S. Pour, M. Ghaemy, Journal of Cleaner Production, Volume 142, January 2017, Page 2973-2984.

[2] C. Michel, C., B. Raveau., Journal. Solid State Chemistry, Volume 90 , December 1991, 296-301.

# Microwave-assisted synthesis of C-doped $TiO_2$ and ZnO as solar-cells electrodes

J. R. Rangel-Mendez[1*], J. Matos[2], L. F. Cházaro-Ruiz[1], A. C. González-Castillo[1], G. Barrios-Yáñez[1]

[1] Instituto Potosino de Investigación Científica y Tecnológica, San Luis Potosí, Mexico
[2] Biomaterials Department, UDT, University of Concepcion, Concepcion, Chile
*rene@ipicyt.edu.mx

## INTRODUCTION

Sunlight is an attractive energy source, clean, and renewable. Some disadvantages of commercial solar cells are high fabrication cost and emission of environmentally unfriendly chemicals during fabrication.

The report on a dye-sensitized solar cell (DSSC) in 1991 by O'Regan and Grätzel is the starting point of a new generation of solar cells. The photosensitization of photoactive semiconductors such as $TiO_2$ and ZnO requires a visible light-induced electron or hole injection into conduction band or valence band, respectively [1]. This is strongly affected by the energy band gap ($E_{bg}$) of the semiconductor; in this regard, it has been reported that carbon- or nitrogen-doped semiconductors induced a red-shift in the $E_{bg}$ and accordingly an increase in the photoefficiency.

Few works have been reported using microwave-assisted hydrothermal synthesis to obtain C-doped $TiO_2$-based materials [2]; microwave-assisted processes permit to reduce the time of reaction and more importantly a better control of the morphology and texture of the materials. In this sense, a systematic study of the influence of the temperature and time of reaction in the microwave-assisted solvothermal synthesis of C-doped $TiO_2$ and ZnO hybrid spheres, using saccharose as carbon source, was performed.

## MATERIALS AND METHODS

In a typical synthesis of C-doped $TiO_2$ and ZnO, 0.5 g of titanium (IV) isopropoxide (97%) or zinc acetate (99.99%), and 0.5 g of saccharose were dissolved in 9 mL of ethanol. The resulting solution was exposed to microwave irradiation at a selected reaction temperature (120°C and 150°C) during a specified time (30 and 150 min for each temperature). The resulting solid was filtered, washed several times with ethanol and dried. Finally, the samples were calcined at 550°C for 5h. All samples were denoted as $TiO_2$-C and ZnO-C followed with the assigned temperature and time of synthesis.

The applied characterization techniques were: SEM, XRD, DR/UV-Vis, XPS, BET and electrochemical characterization.

## RESULTS AND DISCUSSION

Figure 1 shows the SEM micrographs of both materials synthetized at 120°C by 30 min and 150°C by 150 min. As can be seen $TiO_2$ based materials (Fig. 1A and 1B) present an aggregate and spherical morphology with a mean particle size of $0.32 \pm 0.06$ μm for $TiO_2$-C-120-30 and $0.38 \pm 0.10$ μm for $TiO_2$-C-150-150. It can be seen clearly the coalescence interface between the spheres (Fig. 1A and 1B) which is indicative of the coalescence mechanism for the formation of the $TiO_2$ based materials. It is important to remark the high mechanical and thermal stability of the $TiO_2$ based samples because their morphology was not altered by the modification of the synthesis parameters. On the other hand, it is clear that the ZnO-C-120-30 sample (Fig. 1C) present a spiral shape, which is formed by nanoparticles with mean diameter of $0.10 \pm 0.07$ μm. While, the ZnO-C-

150-150 sample has an average particle size of 0.97 ± 0.23 μm (Fig. 1D) and is clearly nanostructured by aggregated of nanoparticles with a mean size of 0.18 ± 0.04 μm. In addition, this sample has empty spaces between nanoparticles forming hollow spheres of ZnO-C.

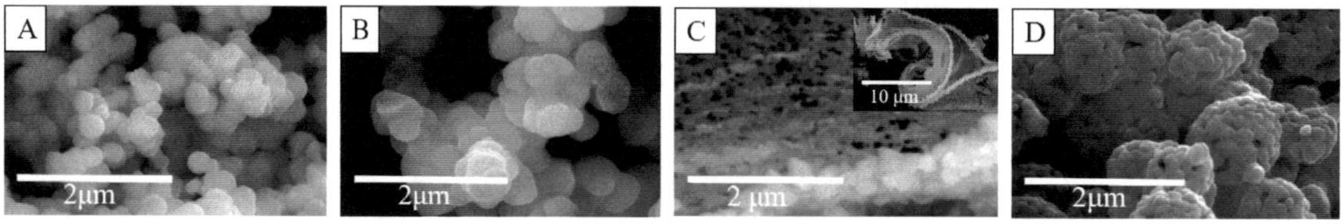


**Fig. 1. SEM micrographs of $TiO_2$-C-120-30 (A), $TiO_2$-C-150-150 (B), ZnO-C-120-30 (C) and ZnO-C-150-150 (D).**

The DR/UV-Vis spectra showed the typical broad absorption band, ascribed to the charge transfer process from $O_2^-$ to $Ti^{4+}$ or $O_2^-$ to $Zn^{2+}$ responsible for the $E_{bg}$ of the semiconductors [3]. As can be observed in Fig. 2A, the calcined $TiO_2$ based samples present a red-shift to the visible range (3.04-3.05 eV) in comparison with the P25-$TiO_2$ (3.26 eV). This is due to the carbon doping within the framework of $TiO_2$ indicating the formation of quantum-dots. Contrary to $TiO_2$ based samples, it can be seen that the calcined ZnO based samples (Fig. 2B) present a lightly red-shift to the visible-light range (3.13-3.18eV) in contrast with the commercial ZnO (3.27 eV). It is clear that the modification of synthesis parameters do not make a remarkable change in the samples $E_{bg}$.

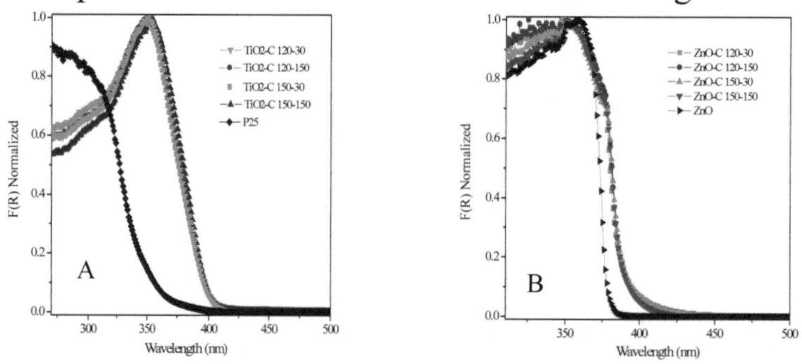


**Fig. 2. UV-Vis/DR spectra plotted as the Kubelka-Munk function of the reflectance F(R) of $TiO_2$-C (A) and ZnO-C (B).**

## CONCLUSIONS

The synthesis of C-doped $TiO_2$ and ZnO hybrid materials by solvothermal microwave-assisted technique was achieved. With the Carbon doping, a red-shift to the visible range in both materials was obtained compared to commercial $TiO_2$–P25 and ZnO.


## ACKNOWLEDGMENTS

The authors appreciate the financial support of project CONACYT-SENER-Sustentabilidad Energética 2014-02 (245467).


## REFERENCES


1. Macyk, W., Szaciłowski, K., Stochel, G., Buchalska, M., Kuncewicz, J., and Łabuz, P. (2010). "Titanium (IV) complexes as direct $TiO_2$ photosensitizers". *Coordination Chemistry Reviews*, Vol.254(21), p.2687-2701

2. Hsu, C. W., Chen, P., and Ting, J. M. (2013). "Microwave-assisted hydrothermal synthesis of $TiO_2$ mesoporous beads having C and/or N doping for use in high efficiency all-plastic flexible dye-sensitized solar cells". *Journal of The Electrochemical Society*, Vol.160(3), p.H160-H165
3. Wang, C., Li, Q., Mao, B., Wang, E., and Tian, C. (2008). "A different chemical route to synthesize ZnO nanoparticles". *Materials Letters*, Vol.62(8), p.1339-1341

# Modified Carbon Nanofiber Films as Interlayers for Lithium Sulfur Batteries

***Liang G***, ***Qin X***, Li Q, Liu M, ***Li B****, Kang F

Presenting author's e-mail: libh@mail.sz.tsinghua.edu.cn

*Graduate School at Shenzhen, Tsinghua University, Shenzhen 518055, China*

## INTRODUCTION

Recently sulfur material has attracted worldwide attention in the Li metal batteries for its high theoretical specific capacity and low cost. However, sulfur cathodes inevitably suffer rapid capacity fade from its poor electronic conductivity and huge volume change in the cycling process. Moreover the intermediates known as the polysulfides appearing during the cycle dissolve easily into the electrolyte and shuttle between the cathode and Li anode, leading to severe active material loss.

In this work, modified carbon nanofiber (CNF) films including $V_2O_5$ decorated, $TiO_2$ coated and cyclized-polyacrylonitrile modified ones were prepared and used as interlayers for Li-S batteries. These modified interlayers have strong absorption for the polysulfides due to the intense electrostatic attraction with polysulfides and they were placed between the cathode and separator to optimize the battery structure design. The Li-S batteries with interlayers mentioned above showed superior cycle performance and excellent rate performance for over 500 cycles.

## MATERIALS AND METHODS

The preparation and structural characterization of the interlayers, as well as the electrochemical performance were reported in the literatures [1,2,3].

## RESULTS AND DISCUSSION

Fig.1 shows the morphology and structure of the $V_2O_5$ modified CNF mat (VCNF). It can be seen that the diameter of origin CNFs was about 300 nm and the modified fibers have a diameter of approximately 400 nm. The layer of $V_2O_5$ was uniformly distributed outside the fiber, making it possible to trap the polysulfides and get easy access to the electronic contact. As shown in the schematic picture, the VCNF could also work as a high potential dam to restrain the self-discharge behavior in the Li-S batteries, which means that Li-S batteries with VCNF interlayer can be stored and rested for a long time without obvious capacity fading. It is worth mentioned that the electrochemical performance of the batteries with VCNF was significantly improved. Even after 1000 cycle testing, the battery with VCNF still delivered a capacity of 700 mAh $g^{-1}$ due to the strong chemical absorption, while the one with CNF interlayer only had the value of about 280 mAh $g^{-1}$.

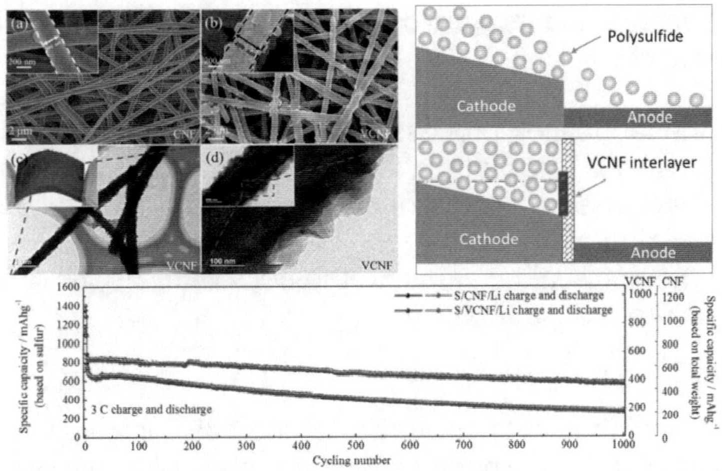


**Fig.1.** The structure of $V_2O_5$ decorated CNF interlayer and battery performance.

Fig.2 shows the schematic of battery structure and working principle of the batteries with $TiO_2$ coated CNF interlayer (TCNF). In this work, we fabricated ultrafine $TiO_2$ nanoparticle modified carbon nanofiber simply through a "dip and dry" method. As a result, not only the well-adsorption of the polysulfides could obtained, but also the fast electronic transfer could be secured at the same time. The battery with TCNF interlayer performed superior electrochemical behavior, which had a capacity retention of 77.78% even after 500 cycles. This simple and low-cost method is very promising in the large scale production.

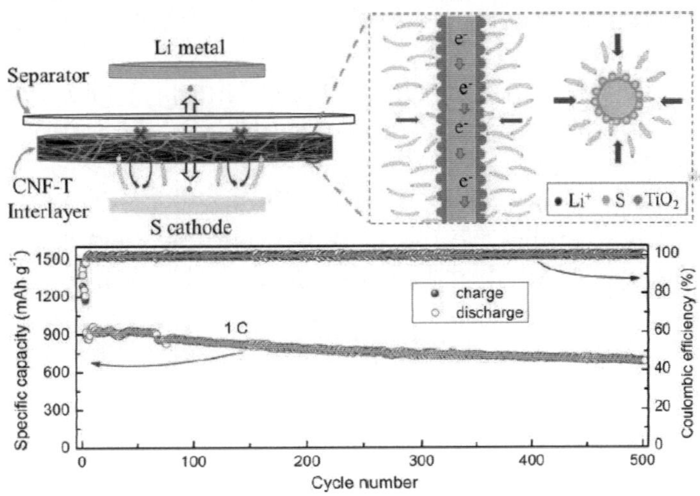


**Fig.2.** The schematic of the $TiO_2$ coated CNF and battery performance.

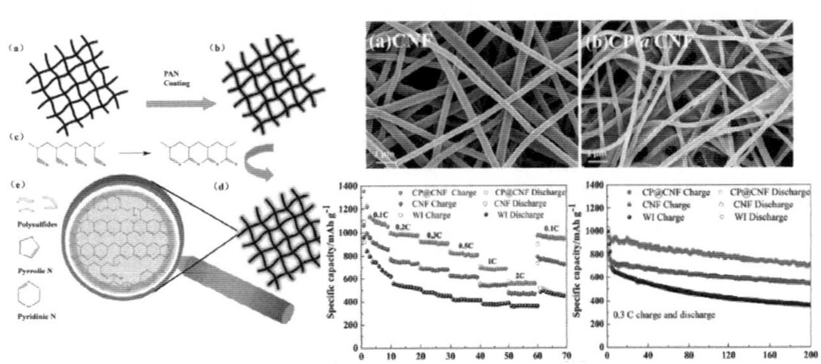

**Fig.3.** The schematic and morphology of the cycled-polyacrylonitrile coated CNF and battery performance.

Fig.3 shows the fabrication process of the cycled-polyacrylonitrile (PAN) coated CNF (CPCNF). PAN layer was cast by a facile dip-coating and during the thermal cyclization treatment there occurred many nitrogen-doped groups in the composite mat, such as the pyridinic nitrogen and pyrrolic nitrogen. These functional groups have strong electrostatic attraction with the polysulfides, making the cells with CPCNF interlayer own a better cycle performance and rate capability.

## CONCLUSIONS

In order to prevent the shuttle behavior of the polysulfides, we designed and fabricated a series of functional material modified CNF interlayer for the Li-S batteries. The CNF network host as a second current collector, providing fast electron transfer and decreasing the battery internal resistance. Different coating layers was designed. For VCNF, it is notable that the self-discharge behavior was effectively suppressed. For TCNF and CPCNF, the fabrication process and condition were simple and facile along with a low cost. Li-S batteries with these interlayers could perform satisfactory cycle stability and rate capability.

## REFERENCES


1. Liu M., Li Q., Qin X., et al. (2017). "Suppressing self-discharge and shuttle effect of lithium–sulfur batteries with $V_2O_5$-decorated carbon nanofiber interlayer", *Small.*, p.1602539.
1. Liang G., Wu J., Qin X., et al. (2016). "Ultrafine $TiO_2$ decorated carbon nanofibers as multifunctional interlayer for high-performance lithium−sulfur battery", *ACS Appl. Mater. Interfaces.*, Vol.8, p.23105−23113.
2. Li Q., Liu M., Qin X., et al. (2016). "Cyclized-polyacrylonitrile modified carbon nanofiber interlayers enabling strong trapping of polysulfides in lithium–sulfur batteries", *J. Mater. Chem. A.*, Vol. 4, p.12973–12980.

# Narrow-diameter Distributed Single-walled Carbon Nanotubes Grown from Carbon Nanorings

Tomohiro Tojo[1*], Ryoji Inada[1], Yoji Sakurai[1]

[1]Department of Electrical and Electronic Information Engineering,
Toyohashi University of Technology, Toyohashi, Aichi, Japan.
*tojo@ee.tut.ac.jp

## INTRODUCTION

Since electronic features of single-walled carbon nanotubes (SWCNTs) strongly depend on the diameter and side wall structure (so-called chirality), the chirality control approach is very important for the promising applications. Although a few physical and chemical approaches have been proposed to obtain single chirality from as-grown materials, it is still difficult to control the electronic feature separation. A new concept without surfactants and metallic catalysts is to directly be synthesized by using carbon nanorings such as cycloparaphenylenes (CPPs) which are the fundamental unit of armchair type SWCNTs with a metallic feature. However, there is still mismatching between inner diameters of SWCNT and CPP during the SWCNT growth process.[1,2] Herein, we propose a new route to synthesize SWCNTs with a defined chirality, using orientated fullerene catalysts encapsulated inside the cavity of CPPs.

## MATERIALS AND METHODS

### Preparation Methods of SWCNTs

Silicon substrates with (100) plane on the surface were cleaned by dipping in a hydrofluoric acid solution (concentration of 46.0-48.0%) at room temperature in order to remove surface oxides, and washed repeatedly with distilled water until the pH became neutral. To prepare atomically flat surfaces on silicon substrates, the treated substrates were dipped into ammonium sulfite solution and washed repeatedly with distilled water, followed by drying under an Ar gas flow. CPP solutions were prepared by dissolving CPPs without and with fullerenes (1:1 by a mole ratio) in toluene. 5 μL of the solution was dropped onto an Si substrate and spin-coated, followed by drying under an Ar gas flow. One more drop of 5 μL of CPP solution without fullerenes was put onto the substrate by the above-mentioned procedures. The prepared substrate was placed into a furnace and heat-treated at approximately 500°C for 15 min under vacuum with focused UV laser irradiation. During the heat treatment, anhydrous ethanol vapor as a carbon source was flowed with an Ar carrier gas into the furnace by a vacuum pressure in order to facilitate the growth of SWCNTs.

## RESULTS AND DISCUSSION

The transmission electron microscopy (TEM) images of the pyrolyzed sample from CPP with $C_{70}$ (SW70 shown in Fig. 1 (a-c)) revealed that amorphous fibers were formed with a diameter ranging from 5 to 30 nm without cylindrical structures attributed to SWCNTs. However, the pyrolyzed CPPs with $C_{76}$ (SW76) and with $C_{84}$ (SW84) were observed in Fig. 1 (d-i) to have both amorphous and cylindrical structures without any obvious difference between their samples even though decomposed molecules remained at the inner and outer nanotubes. The measured diameter distribution of cylindrical structures (Fig. 2) was in the range from 1.5 to 1.7 nm corresponding to the dimeter range of the employed CPPs. Thus, this indicates that the size-selective encapsulation of fullerenes by CPPs determines the growing structures and the cylindrical diameters.

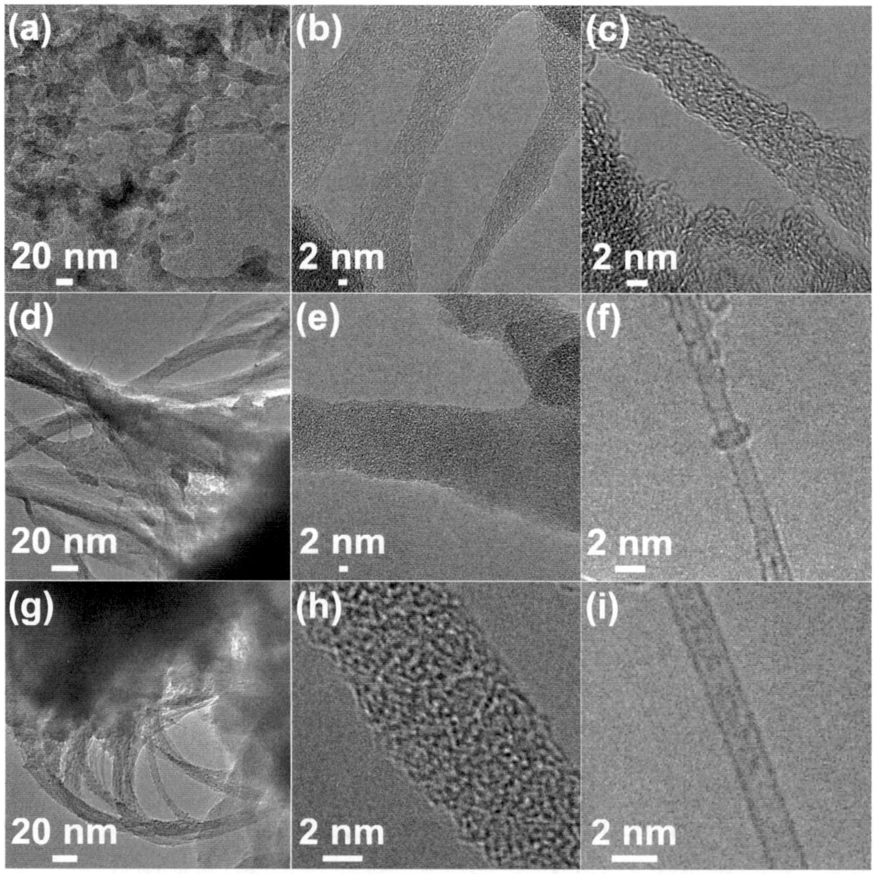

Figure 1. TEM images of SW70 (a-c), SW76 (d-f), and SW84 (g-i).

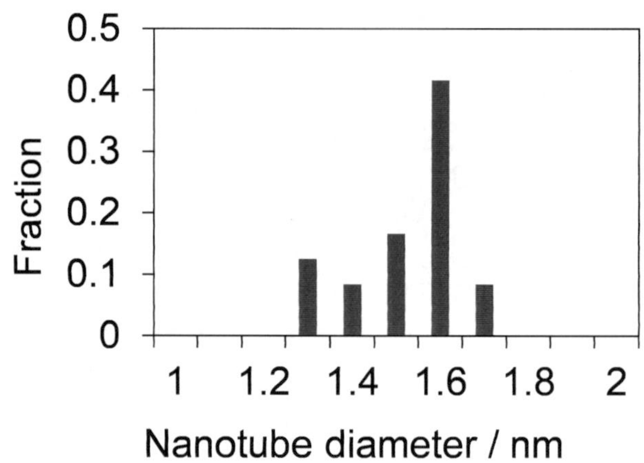

Figure 2. Diameter distribution histogram of SW84 derived from counting the number of cylindrical structures (~25) except amorphous structures.

## CONCLUSIONS

We propose a new approach in which the diameter of fullerenes encapsulated inside the cavity of CPPs determines the diameter distribution of SWCNTs and their cylindrical structures during the SWCNT growth process. This approach could enable design of the electronic features of SWCNTs, thus opening up their possible applications.

## REFERENCES

1. Omachi, H., Nakayama, T., Takahashi, E., Segawa, Y., and Itami, K., (2013). "Initiation of carbon nanotube growth by well-defined carbon nanorings", *Nat. Chem.*, Vol.5(7), pp.572-576.
2. Okada, K., Yagi, A., Segawa, Y., and Itami, K., (2017). "Synthesis and properties of [8]-, [10]-, [12]-, and [16]cyclo-1,4-naphthylenes", *Chem. Sci.*, Vol.8(1), pp.661-667.

# N-doped Carbon Quantum Dots towards Electroluminescence

Jingxia Zheng[1, 2], Yuanfei Ding[1, 2], Yanqin Miao[1, 2], Yongzhen Yang[1, 2*], Xuguang Liu[1, 3*], Husheng Jia[1]

[1]Key Laboratory of Interface Science and Engineering in Advanced Materials, Ministry of Education, Taiyuan 030024, China
[2]Research Center of Advanced Materials Science and Technology, Taiyuan University of Technology, Taiyuan 030024, China
[3]College of Chemistry and Chemical Engineering, Taiyuan University of Technology, Taiyuan 030024, China.
*yyztyut@126.com; liuxuguang@tyut.edu.cn

## INTRODUCTION

Carbon quantum dot-based electroluminescent light emitting diodes (CQD-LEDs), which are the devices based on carbon quantum dots (CQDs) as light emitting layer, have turned out to be a promising candidate for the application in illumination and full-color displays[1,2)]. To improve carrier injection mobility and quatum yield, short-chain passivated and N-doped fluorescent carbon quantum dots (N-CQDs) with graphitic structure were synthesized via a one-step hydrothermal approach using phthalic acid (PA) with benzene ring conjugated structure as carbon source and ethylenediamine (EDA) as nitrogenous additive[3)]. By virtue of the structure characteristics, optical properties and the film-forming ability of N-CQDs, CQD-LEDs were fabricated by solution processing method with N-CQDs as emitting layer, which emit blue light at 6–9 V driving voltage, showing that the N-CQDs have a promising prospect for the application in QD-LEDs.

## MATERIALS AND METHODS

### Material

PA was obtained from Guangfu Technology Development Co., Ltd (Tianjin, China). EDA was purchased from Fuchen Chemical Reagent Factory (Tianjin, China). All chemicals were commercially available and were used without further purification.

### Methods

The schematic illustration of synthesizing N-CQDs for CQD-LEDs is shown in **Fig. 1**. N-CQDs were prepared by one-step hydrothermal treatment using PA as carbon source, EDA as additive and deionized water as solvent. Then solution-processed electroluminescent LED was fabricated using N-CQDs as emitting layer based on the device structure of ITO/poly(ethylenedioxythiophene): polystyrene sulfonate/S-CQDs/1,3,5-tris(N-phenylbenzimidazol-2-yl) benzene/LiF-Al.

## RESULTS AND DISCUSSION

The obtained N-CQDs are structures like crystalline graphite carbon with an average diameter of 5.53 nm, exhibit excitation-dependent PL emission and possess QY of 29.3% in aqueous solutions. The structural analyses of the N-CQDs by Fourier transform infrared and X-ray photoelectron spectroscopy demonstrate that both PA and EDA have engaged in the establishment of N-CQDs. In addition, optical band gap of the N-CQDs was calculated to be 4.26 eV, and, the highest occupied molecular orbital and the lowest unoccupied molecular orbital energy levels of -8.35 and -4.09 eV, respectively, were obtained for the N-CQDs by cyclic voltammetry (**Fig. 2(a)**). The N-CQDs have electron-accepting ability, as proved by electrochemical analysis. The CQD films were nearly uniform and flat when CQD-LEDs were fabricated by using N-CQDs as emitting layer. As shown in **Fig. 2(b), 2(c) and 2(d),** the device emits bright blue light and exhibits good optical stability under the voltage of 6–9 V, a turn-on voltage at about 5.5 V and a maximum luminance of 4.97 cd/m$^2$. The reasons for the low brightness are speculated to be the relatively low QY of N-CQDs and the large energy barrier (3.05 eV) between the emitting layer and the hole injection layer.

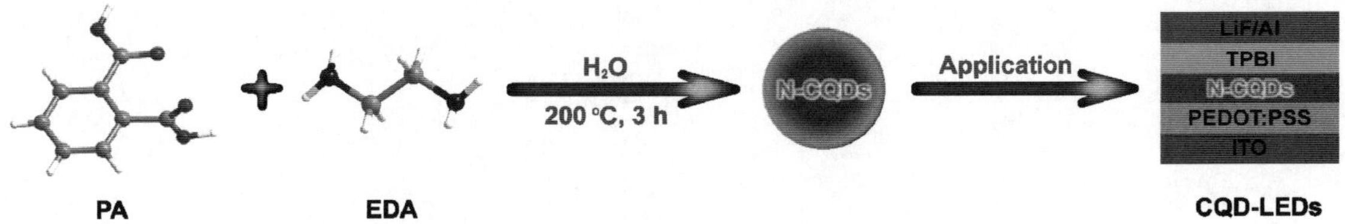

**Fig. 1.** Schematic illustration of synthesizing N-CQDs for CQD-LEDs.

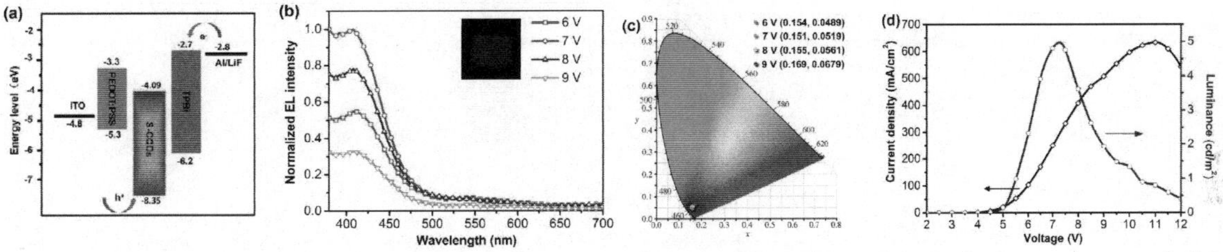

**Fig. 2.** (a) The schematic energy level diagram of the CQD-LEDs. (b) Emission spectra of CQD-LEDs under different working voltage (Inset is the photograph of CQD-LEDs under the working voltage of 7 V). (c) The CIE 1931 chromaticity coordinates of CQD-LEDs under different working voltage. (d) The current density curve and brightness–voltage characteristics of CQD-LEDs.

## CONCLUSIONS

In a word, N-CQDs with graphitic structure were synthesized through a facile hydrothermal method using PA and EDA as the precursor and modifier, respectively. The performance of the CQD-LEDs is anticipated to be further improved by choosing the N-CQDs as emitting layer with high QY and suitable device structure with low energy barrier.


## ACKNOWLEDGEMENTS

This work was supported by the National Natural Science Foundation of China (No. U1610255 and No. U1607120), Shanxi Provincial Key Innovative Research Team in Science and Technology (2015013002-10 and 201605D131045-10) and Shanxi Provincial Key Research and Development Program (201603D111010).


## REFERENCES


1. Sun, Q. J., Wang Y. A., Li, L. S., Wang, D. Y., Zhu, T., Xu, J., Yang, C. H. and Li, Y. F. (2007). "Bright, multicoloured light-emitting diodes based on quantum dots", *Nat. Photonics*, Vol.1, p.717-722
2. Steckel, J. S., Snee, P., Coe-Sullivan, S., Zimmer, J. P., Halpert, J. E., Anikeeva, P., Kim, L.-A., Bulovic V. and Bawendi, M. G. (2006). "Color-saturated green-emitting QD-LEDs", *Angew. Chem. Int. Ed.*, Vol.45, p.5796-5799

3. Zhang, X., Zhang, Y., Wang, Y., Kalytchuk, S., Kershaw, S. V., Wang, Y., Wang, P., Zhang, T., Zhao Y. and Zhang, H. (2013). "Color-switchable electroluminescence of carbon dot light-emitting diodes", *ACS nano*, Vol.7(12), p.11234-11241

# N-doped mesoporous carbons decorated with Pt nanoparticles in hydrotreatment reactions

Cristina Ruiz-García[1*], Francisco Heras[1], Noelia Alonso-Morales[1], Luisa Calvo[1]
Juan J. Rodríguez[1], Miguel A. Gilarranz[1]

[1] Sección de Ingeniería Química, Universidad Autónoma de Madrid, Spain.
*cristina.ruiz@uam.es

## INTRODUCTION

New carbon materials, such as doped carbons have been studied in the last years with the focus in the development of different applications, such as electrochemical devices, $CO_2$ and $H_2$ sorbents, sensors or catalysts. The insertion a doping atom in the carbon matrix modifies chemical and electronic properties, giving an improved performance. It is also well know that materials with a specific morphology can be prepared using a template strategy. By combining these two approaches, it is possible to obtain tailored porous and doped carbons. The objective of this work is to prepare N-doped templated mesoporous carbons and use them as support of Pt nanoparticles to study the influence of N-doping in the activity of the catalysts. The aqueous phase hydrodechlorination (HDC) of 4-chlorophenol (4-CPh) was used as a test reaction to study the influence of N-doping in the catalytic activity.

## MATERIALS AND METHODS

SBA-15 silica template (Sigma-Aldrich), commercial resol resin (Bakelite®PF9934 FL) and 1,10-phenantroline were mixed (1:1:0.1 w) in ethanol. The mixture was placed in a rotatory vacuum evaporator at 100 mbar and 40°C until dryness. The resin infiltrated in the SBA-15 was curated by heating of the dried mixture at 135°C for 1 h in air. The curated resin-silica was treated at 700°C (5ºC $min^{-1}$) in $N_2$ for 1 h. The silica template was removed by washing with 3.5 M NaOH. The templated carbon was washed and oven-dried at 65°C. Pt nanoparticles (NP) were synthetized *in situ* on the templated carbon surface following the colloidal synthesis method described by Lemus et al.[1)], using $H_2PtCl_6{\cdot}6H_2O$ as metal precursor, PVP as capping agent and $NaBH_4$ as reducing agent. The catalysts obtained (2.9% Pt, w. carbon basis) were subjected to hydrothermal treatment at 200°C and 17 bar to remove PVP. Blank catalysts were prepared from templated carbons by incipient wetness impregnation (IWI), calcination at 200°C and reduction at 200°C with $H_2$. The carbons and catalysts were characterized by bulk chemical elemental analysis, $N_2$ adsorption-desorption isotherms at 77 K, transmission electron microscopy (TEM) and X-ray photoelectron spectroscopy (XPS). The reaction experiments were carried out in a batch isotherm reactor at 30 and 70ºC stirred reactor using 150 mL of a 0.1 g/L 4-CPh aqueous solution, 5.8 mg/L of Pt, 50 mL/min of hydrogen feed and 800 rpm of stirring. Activities were calculated from pseudo-first order rate constant obtained from fitting of experimental data for 4-CPh disappearance.

## RESULTS AND DISCUSSION

The TEM images showed that the templated carbons are composed by well-defined arrays of carbon rods formed due to filling of cylindrical pores of starting SBA-15 by resol resin and further pyrolysis. Accordingly, the $N_2$ adsorption-desorption isotherms showed a high contribution of mesoporosity due to intersections of slits and mean size of the pores around 3.4 nm. The XPS spectra showed pyrrolic nitrogen (N-5), imine, amide or amine (N-X) and pyridinic (N-6) as the main contributions in doped carbon. TEM images also showed homogenous distribution of Pt NPs with mean size in the 3.6-3.9 nm range. Therefore, non-doped (CSBA) and N-doped (NCSBA) templated carbons with equivalent pore texture were successfully obtained. The catalysts prepared from templated carbons showed some additional nitrogen resulting from PVP fragments remaining after removal by hydrothermal treatment.

**Table 1.** Characterization of materials and HDC results (5.8 $mg_{Pt}$/L).

| Sample | N (%w) | $S_{BET}$ ($m^2$/g) | $S_{EXT}$ ($m^2$/g) | pore (nm) | Pt size (nm) | activity ($mmol.g_{Pt}$/min) |
|---|---|---|---|---|---|---|
| CSBA | 0.1 | 1273 | 730 | 3.4 | - | - |
| NCSBA | 0.7 | 1390 | 722 | 3.3 | - | - |
| CSBA/Pt-NP | 0.6 | | | | 3.6 | 20-41* |
| NCSBA/Pt-NP | 0.8 | | | | 3.9 | 17-50* |
| CSBA/Pt-IWI | | | | | 4.6 | 17** |
| NCSBA/Pt-IWI | | | | | 2.1 | 25** |

* values in the 30-70ºC range, ** values at 50ºC and 3.4 $mg_{Pt}$/L

HDC experiments using a Pt concentration in the reaction medium of 2.4-10.0 mg/L showed fast conversion of 4-CPh and almost negligible changes in activity, evidencing chemical control. However, for reactions at 10 mgPt/L cyclohexanol was the only product after 60−120 minutes, showing high capability for the hydrogenation of phenol, the primary product of 4-CPh HDC, with NCSBA/Pt-NP exhibit higher hydrogenation ability. However, the activity at 30ºC was slightly higher for CSBA/Pt-NP.

The temperature influence (30−70ºC) was studied using a Pt concentration of 5.8 mg/L. The activation energy in the range was 15±2 and 23±5 $kJmol^{-1}$ for CSBA/Pt-NP and NCSBA/Pt-NP, respectively, showing the influence of carbon doping, a thus achieving higher activity for NCSBA/Pt-NP at 70ºC. The catalysts based on Pt-NPs showed higher activity than the ones obtained by IWI. NCSBA/Pt-IWI catalyst showed higher activity than CSBA/Pt-IWI, indicating that the support the doped support contributes to activity. This effect may be through the dispersion of the nanoparticles during the reduction, although Pt-support interaction can also be relevant. On the contrary, when the Pt NPs were synthesized in the PVP environment, a lower interaction with the support during Pt precursor reduction took place and the doping of the support did not lead to so clear enhanced activity.

## CONCLUSIONS

Pt catalysts supported on nitrogen-doped and non-doped templated carbon with equivalent porosity and structure of the metal phase have been prepared and applied in the HDC of 4-CPh. Catalysts based on templated non-doped carbon showed slightly higher activity at 30ºC, whereas those based on nitrogen-doped carbon showed higher hydrogenation capability and higher activity at 70ºC.

## ACKNOWLEDGEMENT

The authors are grateful to MINECO for support (CTQ2012-32821, CTQ2015-65491-R) and to Hexion Specialty Chemicals Ibérica for phenolic resin samples. C. Ruiz-García acknowledges for PhD grant (BES-2013-066085).

## REFERENCES

1. Lemus, J., Bedia, J., Calvo, L., Simakova, I.L., Murzin, D. Yu, Etzold, B.J.M., Rodriguez, J.J., Gilarranz, M.A. (2016). "Improved synthesis and hydrothermal stability of Pt/C catalysts based on size-controlled nanoparticles" Catal. Sci. Technol., Vol. 6, p. 5196-5206.

# $NH_4FeF_3$/Carbon-Nanosheet Composite as a Potential Anode Material for Li Ion Storage

Minhong Kong[1], Jisheng Zhou[1, 2*], Xiaohong Chen[1], Huaihe Song[1, 2*]

[1]State Key Laboratory of Chemical Resource Engineering, Beijing Key Laboratory of Electrochemical Process and Technology for Materials, Beijing University of Chemical Technology, Beijing, P. R. China
[2] Changzhou Institute of Advanced Materials, Beijing University of Chemical Technology, Jiangsu, P. R. China.
*zhoujs@mail.buct.edu.cn

## INTRODUCTION

Transitional metal fluorides (TMFs) have gained growing attentions as promising electrode materials for lithium ion batteries (LIBs) due to their resource abundance, low-cost and high specific capacities.[1] For example, Li et al. prepared iron fluoride nanowires ($FeF_3$ NWs) by dehydrating the $\alpha$-$FeF_3 \cdot 3H_2O$ precursor, which exhibited a specific capacity of 223 mA h $g^{-1}$ at 50 mA $g^{-1}$ for LIBs.[2] However, these TMFs still suffer from some disadvantages, such as lower conductivity owing to large bond gap of TMFs, and larger volume expansion during the charge/discharge process, which deteriorate cyclic stability and rate-performance, and limit their practical applications. In order to improve the conductivity of TMFs, various carbon materials such as carbon black, carbon nanotube and graphene have been employed to composite with TMFs. Additionally, in order to improve their electrochemical performance for LIBs, it is important to pay special attentions to crystal topologies of TMFs, which have a key effect on their ionic conductivity. Although the design of expanded structures is favourable for improving ionic conductivity, the intrinsic electron conductivity of such open-structured fluorides is still undesirable to obtain high-rate performance. Similarly, electrochemical performance of TMFs with open framework structure should also be enhanced by compositing with carbon nanomaterials. However, the preparation of open framework TMFs/carbon composites is still few reported.

## MATERIALS AND METHODS

### Material

The $NH_4FeF_3$/CNS composites were prepared by in situ co-pyrolysis of ferric acetylacetonate and $NH_4F$. The in situ co-pyrolysis method is a valid method to synthesis new-types of carbon nano-materials. In our previous work, varieties of carbon nano-materials such as carbon-encapsulated $Fe_3C$/Fe nanocrystals[3] and quadrangular carbon nanotubes[4] have been prepared via in situ co-pyrolysis by employing diverse sources and regulating experiment conditions.

## RESULTS AND DISCUSSION

Fig. 1a is the SEM image of the $NH_4FeF_3$/CNS composites. It can be seen that the $NH_4FeF_3$ nanoparticles are anchored on the wave-like carbon nanosheets. And the corresponding TEM image (Fig. 1b) further shows that the diameter of the $NH_4FeF_3$ nanoparticles is in the range from 30 to 100 nm.

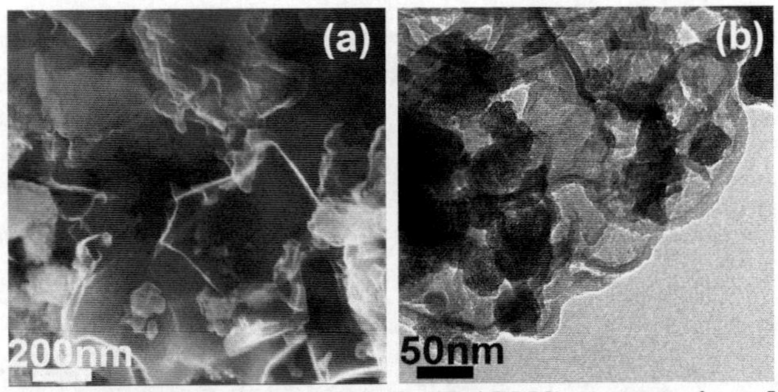


**Fig. 1 (a) SEM and (b) TEM images of the $NH_4FeF_3$/CNS composite obtained at 500 °C for 2 h.**

The XRD pattern of $NH_4FeF_3$/CNS composite (Fig. 2a) shows that the main peaks are indexed well to $NH_4FeF_3$ with cubic perovskite structure according to JCPDS card (PDF#54-1071), and some weak peaks are attributed to $FeCO_3$, indicating trace impurities. The rate capability of the products at various current densities was confirmed in Fig. 2b. At higher current densities of 100, 200, 500, 1000 and 2000 mA $g^{-1}$, the reversible capacities of $NH_4FeF_3$/CNS can reach up to 787, 700, 610, 535 and 463 mA h $g^{-1}$, respectively. After the current density returns to 50 mA $g^{-1}$, the reversible capacity can be kept ca. 883 mA h $g^{-1}$, which is 93% of the first reversible capacity of 953 mA h $g^{-1}$ at 50 mA $g^{-1}$.

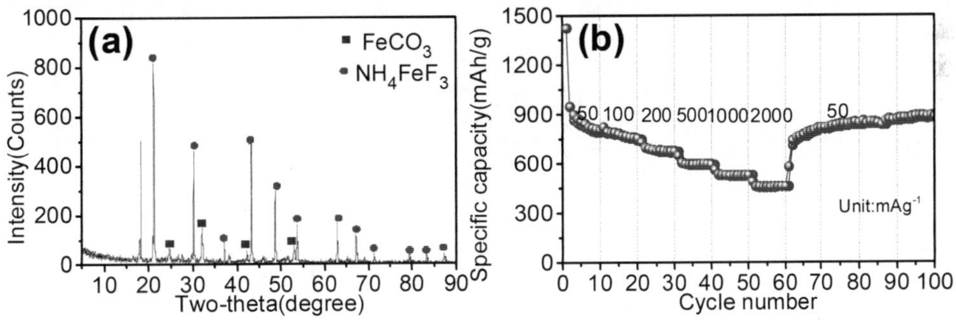


**Fig. 2 (a) XRD pattern and (b) rate performance of $NH_4FeF_3$/CNS composites.**

## CONCLUSIONS

As anode materials for LIBs, $NH_4FeF_3$/CNS composite exhibits superior rate-performance and excellent cyclic stability. Therefore, this work opens the door to explore other transitional metal fluorides to be applied in other fields such as supercapacitors, catalysts, etc.

## REFERENCES

1. Ding, C., Zeng, Y. -W., Cao, L. -L., Zhao L. -F. and Zhang, Y. (2016). "Hierarchically porous $Fe_3O_4$/C nanocomposite microspheres via a CO2 bubble-templated hydrothermal approach as high-rate and highcapacity anode materials for lithium-ion batteri", J. Mater.Chem. A, Vol.4(16), p.5898-5908

2. Li, L. -S., Meng, F and Jin, S. (2012). "High-capacity lithium-ion battery conversion cathodes based on iron fluoride nanowires and insights into the conversion mechanism", *Nano Lett.*, Vol.12(11), p.6030-6037

3. Zhou, J. -S., Song, H. -H., Chen, X. -H., Zhi, L. -J., Huo, J. -P and Cheng, B. (2009). "Oxidation Conversion of Carbon-Encapsulated Metal Nanoparticles to Hollow Nanoparticles", *Chem. Mater.*, Vol. 21(15), p. 3730-3737

4. Zhou, J. -S., Song, H. -H., Fu, B.-H., Wu, B. and Chen, X. -H. (2010). "Synthesis and high-rate capability of quadrangular carbon nanotubes with one open end as anode materials for lithium-ion batteries", *J. Mater. Chem.*, Vol.20(14), p.2794-2800

# Nitrogen doped carbon aerogel - graphene composite materials for electrocatalysis

Krisztina László[1]*, Balázs Nagy[1], Imre Bertóti[2], Miklós Mohai[2,] István Bakos[2],

[1] Budapest University of Technology and Economics, Budapest, Hungary
[2] Research Centre for Natural Sciences, Hungarian Academy of Sciences, Budapest, Hungary.
*klaszlo@mail.bme.hu

## INTRODUCTION

Nitrogen is the most versatile heteroatom in carbon matrices. It affects both the physical and the chemical behaviour of carbons already at low concentration (<1.5%) by influencing the electron distribution and thus the electrical, hydrophilic/hydrophobic and catalytic properties. Nitrogen doped carbon materials have a great potential in energy conversation and storage e.g., as catalyst in the oxygen reduction reaction (ORR) of fuel cells [1]. The electrical conductivity of the carbon materials is crucial in these applications. Incorporation of graphene (GR) into the porous carbon matrix can further improve the electrical conductivity without deteriorating their beneficial properties.

## MATERIALS AND METHODS

### Material

In this work nitrogen doped carbon aerogel - graphene composite materials were prepared from resorcinol (R) – melamine (M) – formaldehyde (F) copolymers by aqueous sol-gel methodology [2]. Graphene was incorporated as graphene-oxide (GO) suspension during the polymerization process (Table 1). The polymer hydrogels were dried by supercritical $CO_2$ and converted to carbon by heat treatment in inert atmosphere. Samples prepared without melamine were used for comparison.

**Table 1 Composition of precursor sol and sample abbreviations**

| Components | Abbrevation | In precursor | |
|---|---|---|---|
| | | M/R molar ratio | R/GO mass ratio |
| R+F | RF | 0 | - |
| R+F + 10 % GO | RF + GO | 0 | 10 |
| R+M+F_20 | RMF | 0.2 | - |
| R+M+F_20 + 10 % GO | RMF + GO | 0.2 | 10 |

The morphology of the carbon aerogels obtained was characterized by low temperature nitrogen adsorption/desorption isotherms and scanning electron microscopy (SEM). The surface chemical composition was determined by X-ray photoelectron spectroscopy (XPS). The catalytic performance in ORR was followed by cyclic and linear sweep voltammetry (CV, LSV).

## RESULTS AND DISCUSSION

As Fig. 1 a-b shows, the GO layers are evenly distributed in the dry polymer precursor which seem to behave as “support” for the polymer spheres. Their effect on the porosity is obvious from the pore size distribution deduced from $N_2$ adsorption measurements (Fig. 1c). The effect of the GO on the surface chemistry (XPS) in the corresponding samples is not significant.

The electrical conductivity (DC) is the best in the RMF+GO sample (Table 2). The positive effect of both the melamine and the GO is clear.

a b c

**Fig. 1 SEM images of the RF (a) and RF_GO polymer aerogel (b) and the pore size distribution of the polymer aerogel samples (c)**

**Table 2 .Electrical (DC) conductivity**

| Sample | Conductivity (S/cm) |
|---|---|
| RF | 0.4 |
| RF + GO | 0.7 |
| RMF | 0.9 |
| RMF +GO | 1.4 |

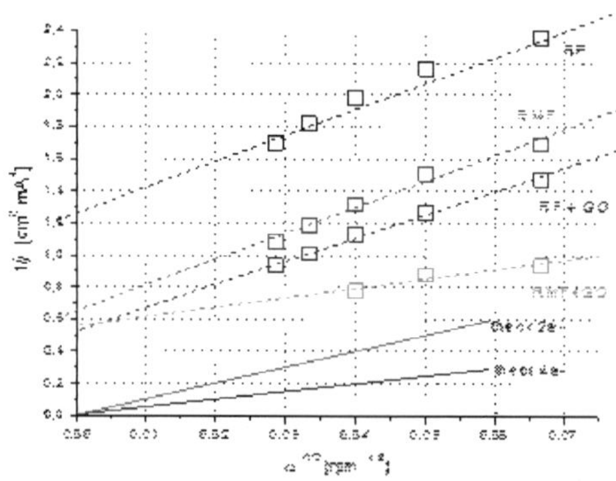

**Fig. 2 Koutecky-Levich plot of the LSV results**

The ORR activity was tested by linear sweep voltammetry (LSV, 0.5 M aq. $H_2SO_4$ solution, saturated with oxygen). The cathodic current was measured and evaluated with the Koutecky-Levich equation (Figure 2). The corresponding plots of the 2 and 4 electron processes are shown for comparison.

## CONCLUSIONS

GO doped carbon aerogels without and with N content were synthesised and tested in electrocatalysisi. The influence of the evenly distributed GO on the polymer gel morphology is the most obvious in the micropore region. The carbon aerogel with combined N and GO content enhance the DC conductivity (ca 350 %). All the samples are more active in electrocatalysis than the carbonized RF itself and exhibit a two-electron process except the carbonized RMF + GO sample, where the mechanism electrocatalytical involves 4 electrons.


## ACKNOWLEDGEMENT

The authors are grateful to the Institut Laue-Langevin and ISIS Neutron and Muon Source facilities access to the instruments. This research was supported by the Hungarian Scientific Research Fund K109558. The work is part of the EU project FP7-PEOPLE-2010-IRSES-269267 and the Nanoporous and Nanostructured Materials for Medical Applications (Nanomed: H2020-MSCA-RISE-2016, project:734641).


## REFERENCES

1. Guo, D., Shibuya, R., Akiba, C., Saji, S., Kondo, T., Nakamura, J. (2016). "Active sites of nitrogen-doped carbon materials for oxygen reduction reaction clarified using model catalysts", *Science* Vol.351, p.361–365
2. Nagy, B., Villar-Rodil, S., Tascón, J. M. D., Bakos, I., László. K. (2016). "Nitrogen doped mesoporous carbon aerogels and implications for electrocatalysisˇ, *Micropor. Mesopor. Mater.,* Vol.230, p.135-144

***Some figures and text in this paper have been edited to make it more readable***

# NITROGEN-DOPED CARBON MATERIALS ANALYZED BY X-RAY PHOTOELECTRON SPECTROSCOPY

Haruki Tanaka[1], Yasuhiro Yamada[1*], Satoshi Sato[1], and Shingo Kubo[2]

[1](*Graduate School of Engineering, Chiba University, 1-33 Yayoi, Inage, Chiba, 263-8522, Japan*)
[2](*Division of Instrumental Analysis, Natural Science Center for Research and Education, Kagoshima University, 1-21-40 Korimoto, Kagoshima, 890-0065, Japan*)
*(*y-yamada@faculty.chiba-u.jp*)

## INTRODUCTION

Types of nitrogen-containing functional groups play a crucial role to improve those properties, but the detail analysis of the structure of nitrogen-doped carbon materials has not been conducted. X-ray photoelectron spectroscopy (XPS) has been well utilized to analyze nitrogen-doped carbon materials. Owing to the efforts of many researchers, the assignments of nitrogen-containing functional groups have been gradually revealed. However, some problems still exist on the assignment of nitrogen-containing functional groups. Nitrogen-containing functional groups have been classified into pyridinic, pyrrolic, amine ($NH_2$), and either graphitic/quaternary nitrogen. However, other similar functional groups can exist in nitrogen-doped carbon materials. In this work, nitrogen-doped carbon materials with various amine groups were constructed as model structures. C1s and N1s XPS spectra of those structures were simulated by density functional theory calculation and N1s XPS peak shifts of various amine groups depending on surrounding environment were revealed.

## MATERIALS AND METHODS

Acridine tetramer having a structure similar to nitrogen-doped graphene nanoribbons was used as a basic structure to analyze the peak positions of different types of amine. C1s and N1s XPS spectra of two types of model structures such as nitrogen-doped graphene nanoribbon-like structure containing five types of amine (defined as amine 1-5 (**Fig. 1**)) were simulated using density functional theory (DFT) calculation (B3LYP/6-31g(d) integral=grid=ultrafine pop=full gfprint, Gaussian 09). Scaling factors previously reported by our group were applied to adjust simulated spectra to actual spectra.[1)]

## RESULTS AND DISCUSSION

**Fig. 1** shows simulated N1s XPS shifts of nitrogen-doped graphene nanoribbon like structures with different nitrogen functional groups. The peak positions of Amine 1 for each model structure are almost constant (399.3 eV), which was consistent with the results of our previous study.[1)] Peak positions of Amine 1-5 were 399.3, 399.8, 399.9, 400.4, and 400.8 eV, respectively. The difference of peak tops between pyridinic and amine groups increased from 0.5 to 2.2 eV as the density of six-membered ring around amine-nitrogen increased. The peak position of pyridinic groups shifted depending on the surrounding bonding state of nitrogen.

The peak positions of these amines are summarized in **Table 1**. Clear differences in N1s peak shifts among Amine 1-5 were obtained. We could reveal N1s peak shifts of various amine groups depending on surrounding environment.

***Some figures and text in this paper have been edited to make it more readable***

Table 1 N1s peak shifts of amine-containing carbon materials.*nanoribbon.

| | Structures | Simulated Binding energy (eV) (Previous study) |
|---|---|---|
| Amine 1 ($sp^2C-NH_2$) | [redacted] | 399.3 (399.3 - 4)[1] |
| Amine 2 (Tertiary amine) | [redacted] | 399.8 (399.7)[2] |
| Amine 3 | 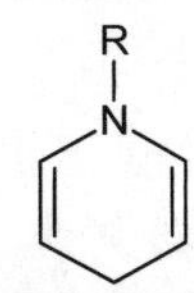 | 399.9 |
| Amine 4 (Valley amine) | [redacted] | 400.4 (400.6 - 401.1)[1] |
| Amine 5 (Basalamine) | 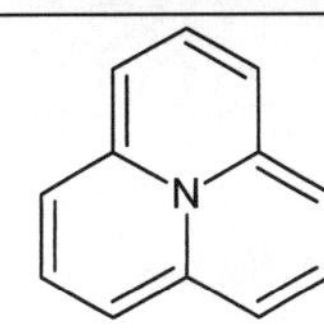 | 400.8 (400.3 - 401.1)[1] |

*Peak top of C=C was adjusted 284.3 eV and the adjusted C1s shift was applied to N1s to adjust N1s shift.

## REFERENCES


1. Yamada, Y., et al. (2014). "Nitrogen-containing graphene analyzed by X-ray photoelectron spectroscopy", *Carbon*, Vol. 70, p.59-74

2. R. J. J. Jansen, et al. (1994). "XPS of nitrogen-containing functional groups on activated carbon", *Carbon*, Vol. 33, p.1021-1027

# Two-dimensional Nitrogen-doped Carbon Nanosheets Supported $NiSe_2$ Nanoparticles as Anode Material for Enhanced Sodium Storage

Sitong Liu[1], Jisheng Zhou[1, 2*], Xiaohong Chen[1], Huaihe Song[1, 2*]

[1] State Key Laboratory of Chemical Resource Engineering, Beijing Key Laboratory of Electrochemical Process and Technology for Materials, Beijing University of Chemical Technology, Beijing, P. R. China

[2] Changzhou Institute of Advanced Materials, Beijing University of Chemical Technology, Jiangsu, P. R. China

zhoujs@mail.buct.edu.cn; songhh@mail.buct.edu.cn

## INTRODUCTION

Sodium ion batteries (SIBs) have attracted significant attention for next-generation electrical energy storage systems due to low cost and abundant sodium resources[1)]. Nevertheless, the lack of suitable electrode material is one of the biggest bottlenecks for the successful development of SIBs. In this regard, transitional metal selenides (TMDs) have gained increasing interest because of the high theoretical capacity and intrinsic high conductivity[2)]. Among various TMDs materials, $NiSe_2$ processed with unique properties is a promising electrode material[3)]. Unfortunately, similar like other transitional metal compounds, $NiSe_2$ also suffers from large volume changes during charge/discharge process, which leads to rapid capacity fading and poor rate capability. To overcome these disadvantages, an effective strategy is to decorate TMDs with conductive carbon matrixes, and the electrochemical performance can be further enhanced by the introduction of heteroatoms. In this work, a novel N-doped carbon nanosheets supported $NiSe_2$ nanoparticles (NCNs-$NiSe_2$) was synthesized by a simple pyrolysis of hexamine (HMT)/$Ni(NO_3)_2$ mixture followed by a simple selenization process. In the as-prepared NCNs-$NiSe_2$, the $NiSe_2$ nanoparticles were uniformly distributed in the two-dimensional nitrogen doped carbon matrix. The N-doped carbon matrix can significantly improve the conductivity of the composite and buffer the volume expansion of $NiSe_2$ nanoparticles during the charge/discharge process. When served as anode material for SIBs, the novel electrode exhibited a high reversible capacity of 410 mAh $g^{-1}$ at a current density of 1 A $g^{-1}$ and still owned a capacity of 310 mAh $g^{-1}$ after 100 cycles. The excellent sodium storage capability can be ascribed to the synergistic effects of nitrogen doped carbon nanosheets and $NiSe_2$ nanoparticles.

## MATERIALS AND METHODS

An aqueous solution of $Ni(NO_3)_2 \cdot 6H_2O$ was added dropwise to a stirred solution of HMT with a molar ratio of 1:2 and immediately the green precipitates were generated. Afterwards, the HMT/$Ni(NO_3)_2$ mixture can be obtained by centrifugation and washing by alcohol several times to remove unreactive sources. Subsequently, the mixture was annealed at 500 °C for 2 h and followed by a selenization process at 400 °C for 5 h.

## RESULTS AND DISCUSSION

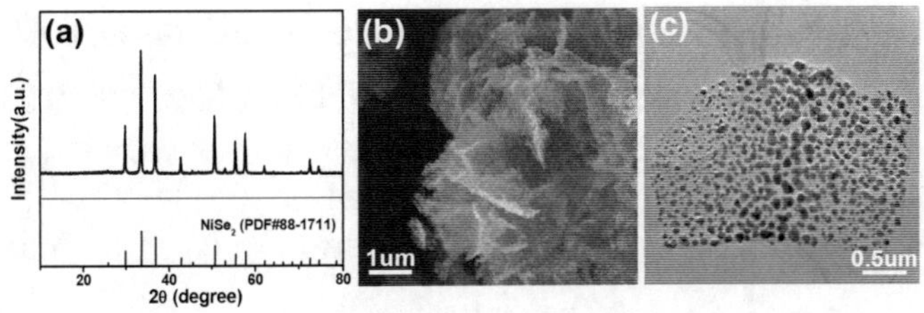


**Fig. 1.** (a) XRD pattern of NCNs-$NiSe_2$; (b) SEM and (c) TEM images of NCNs-$NiSe_2$.

The composition and structure of as-synthesized NCNs-$NiSe_2$ was first investigated by XRD (Fig. 1a). The diffraction peaks of NCNs-$NiSe_2$ are well consistent with the cubic $NiSe_2$ standard pattern (PDF# 88-1711). The SEM (Fig. 1b) and TEM (Fig. 1c) images show that $NiSe_2$ nanoparticles with a diameter of 100-200 nm are uniformly distributed in the carbon nanosheets matrix.

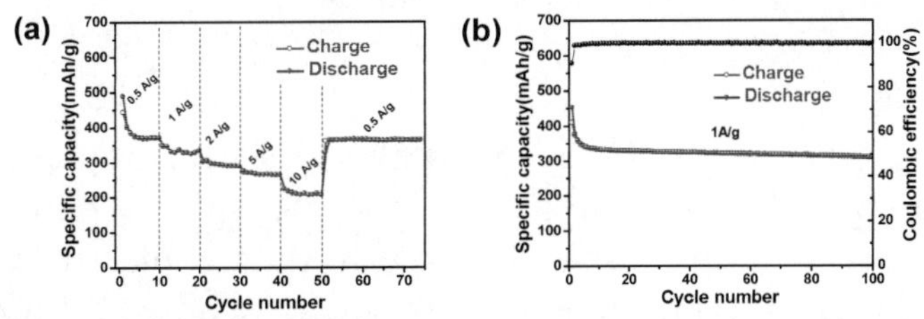


**Fig. 2.** (a) Rate capability of NCNs-$NiSe_2$ electrode; (b) cycling performance of NCNs-$NiSe_2$ electrode at 1 A $g^{-1}$.

Fig. 2a shows the rate performance of NCNs-$NiSe_2$ measured in the voltage range of 0.1-2.5 V. The NCNs-$NiSe_2$ exhibits a reversible capacity of 410 mAh $g^{-1}$ at 0.5 A $g^{-1}$. Even at 10 A $g^{-1}$, the electrode still retains a high capacity of 210 mAh $g^{-1}$, which is ca. 51% of the reversible capacity at 0.5 A $g^{-1}$. Fig. 2b shows the cycling performance of NCNs-$NiSe_2$ at 1 A $g^{-1}$, which can deliver a capacity of 310 mAh $g^{-1}$ after 100 cycles.

## CONCLUSIONS

Novel NCNs-$NiSe_2$ was synthesized by a simple pyrolysis of HMT/$Ni(NO3)_2$ mixture followed by a simple selenization process. Served as anode material for SIBs, the NCNs-$NiSe_2$ exhibits a superior high-rate Na-ion storage. The excellent sodium storage capability can be ascribed to the synergistic effects of nitrogen doped carbon nanosheets and $NiSe_2$ nanoparticles.

## REFERENCES

1. Slater MD, Kim D, Lee E, Johnson CS. (2013). "Sodium-ion batteries", *Adv. Funct. Mater.*, Vol.23(8), p.947-58.

2. Lu T, Dong S, Zhang C, Zhang L, Cui G. (2017). "Fabrication of transition metal selenides and their applications in energy storage", *Coord. Chem. Rev.*, 332, p.75-99.

3. Cho JS, Lee SY, Kang YC. (2016). "First introduction of $NiSe_2$ to anode material for sodium-ion batteries: a hybrid of graphene-wrapped $NiSe_2$/C porous nanofiber", *Sci. Rep.*, 6, p.23338.

# Nitrogen-doped carbon nanotubes derived from metal–organic framework as an efficient ORR electrocatalyst

Rui Li[1], Xuzhen Wang[1,2,*], Yanfeng Dong[2], Lan Yang[1], Jieshan Qiu[2]

1. School of Chemistry, Dalian University of Technology, Dalian, China.
2. Liaoning Key Lab for Energy Materials and Chemical Engineering, Faculty of Chemical, Environmental & Biological Science and Technology, Dalian University of Technology, Dalian, China.
Presenting author's e-mail: xzwang@dlut.edu.cn

## INTRODUCTION

Oxygen reduction reaction (ORR) electrocatalysis is very important for many energy storage and conversion technologies.[1] Zeolitic imidazolate frameworks (ZIFs), a novel material, can be used as ideal precursors for the synthesis of high-performance electrocatalysts [2] with uniform nitrogen doping *via* facile pyrolysis. Unfortunately, the ZIF-67 nanobulks are mostly in low surface area and large size particles, leading to unsatisfactory electrochemical activity. By a template induced strategy, the hierarchical composite of ultrathin wall N-doped carbon nanotubes decorated with cobalt nanoparticles (Co@C-NCNTs) have been constructed from ZIF-67 and halloysite nanotubes (HNTs). Benefited from one-dimension (1D) nanotubes owning more exposed active surface area as well as the convenient channels for transport of electrons and reactants, the Co@C-NCNT hybrid exhibits comparable catalytic activity and outperformed durability for ORR, compared with intrinsic nanoparticles by direct carbonization of ZIF-67 crystals. Additionally, the hybrid shows superior methanol tolerance both in alkaline and acidic solution when benchmarked with a commercial Pt/C electrocatalyst. The synthesis method for structure-control electrocatalysts would open a new way on preparing promising cathode catalyst candidates for fuel cell applications.

## MATERIALS AND METHODS

### Material

HNTs powder was purchased from Yanbo minerals processing plant. Cobalt nitrate hexahydrate and hydrofluoric acid were obtained from Kermel Chemical Reagent Company (Tianjing, China). Dimethyl-imidazole (DMI) was obtained from Guangfu Fine Chemical Reaearch Institute (Tianjing, China).

### Methods

*Synthesis of ZIF-67 derived Co@C-NCNT.* 0.78g of HNTs and $Co(NO_3)_2{\cdot}6H_2O$ (2.91 g) was dispersed in 80 ml of anhydrous ethanol under ultrasonic for 30 min. The solution of DMI (3.284 g) with 80 mL methanol was added to the above solution, then standed at RT for 24 h. After washed three times with anhydrous ethanol and dried at 70 $^{o}$C overnight, the as-prepared HNT@ZIF-67 was annealed at 800$^{o}$C for 2 h under flowing $N_2$ atmosphere, with a ramp of 5$^{o}$C min$^{-1}$. At last, the carbonized samples were immersed in HF solution for 12 h and washed thoroughly with lots of DI water.

## RESULTS AND DISCUSSION

Static electricity induced adsorption on the surface of HNTs leads to the confined and *in situ* growth of thin layer of ZIF-67. TEM image of HNT@ZIF-67 (Fig.1a) shows that the well-dispersed ZIF-67 nanocrystals with sizes of 20~30 nm were coated on tubular HNTs. After transformed into Co@C-NCNT (Fig.1b), the typical tubular morphology of the hybrid is kept while with shrinked size in comparison with HNT@ZIF-67 composites. Subsequent carbonization and template removal give rise to ultrathin wall NCNTs and large inner cavity (Fig.1b, and the inset image).

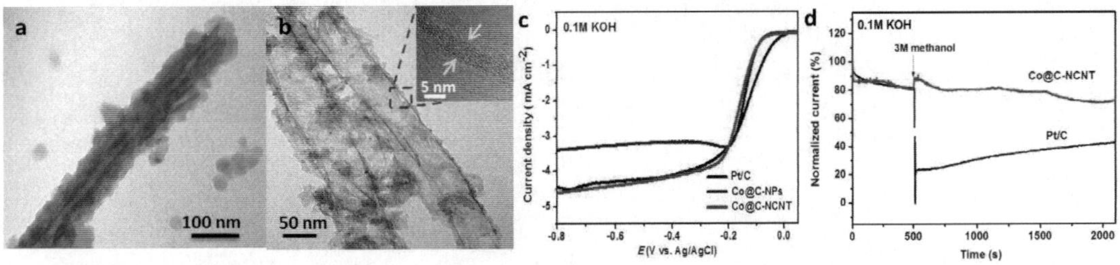


**Figure 1.** TEM images of (a) HNT@ZIF-67 hybrid and (b) Co@C-NCNT derived from HNT@ZIF-67. The inset is HR-TEM image of the wall of Co@C-NCNT. (c) LSVs of Pt/C, Co@C-NCNT and Co/C-NPs electrodes at a rotation rate of 1600 rpm. (d) The durability of Co@C-NCNT and Pt/C electrodes upon addition of 3 M methanol into $O_2$-satuated KOH at -0.6 V (added at 500 s).

Linear sweep voltammograms (LSVs) investigated by a rotating disk electrode (RDE) in $O_2$-saturated 0.1 M KOH electrolyte at a scan rate of 10 $mV \cdot s^{-1}$ and at a rotation rate of 1600 rpm would gain insight into the electrocatalytic activity of Co@C-NCNT towards the ORR. As shown in Fig.1c, the LSV polarization curve of Co@C-NCNT displays a close onset potential (-0.104 V) with Co@C-NPs (-0.099 V), and only 60 mV negative shift compared to the commercial Pt/C (20 wt%) catalyst. In particular, the half-wave potential ($E_{1/2}$) of Co@C-NCNT was at -0.162 mV, which is approaching to Pt/C (-0.134 mV). In addition, the diffusion-limiting current of Co@C-NCNT (4.42 mA $cm^{-2}$) was higher against its counterparts Pt/C (4.28 mA $cm^{-2}$) and Co@C-NPs (3.47 mA $cm^{-2}$). The current–time (*I–t*) technique is used to test the tolerance of Co@C-NCNT and Pt/C catalyst against methanol. As shown in Fig.1d, the current density for Co@C-NCNT has a slight oscillation and recovered quickly after adding 3 M methanol, while a sharp decrease for Pt/C. The results demonstrate that Co@C-NCNT hybrid has appealing properties against crossover effect, making it a promising catalyst for direct methanol fuel cells.

## CONCLUSIONS

An effective approach for the preparation of N-doped carbon nanotubes derived from ZIF-67 *via* a template-induced strategy as a new efficient ORR electrocatalyst has been demonstrated. It was found that most active Co@C-NCNT catalyst exhibits comparable ORR activity with its counterpart commercial Pt/C in terms of close onset potential (-0.104 V), high current density up to -4.42 mA $cm^{-2}$. Our study provides a new design strategy for preparing metal–organic frameworks derived one-dimensional Co-N-C functional materials for energy storage and conversion application.

## REFERENCES


1. Du, R., Zhang, N., Zhu, J., Wang, Y., Qu, L., Hou, Y. and Zhang, J. (2015). "Nitrogen-Doped Carbon Nanotube Aerogels for High-Performance ORR Catalysts", *Small.*,11(32), 3903-3908.

2. Zhong, H. X., Wang, J., Zhang, Y. W., Xu, D., Zhang, Y. F. and Zhang, X. B. (2014). "ZIF-8 derived graphene-based nitrogen-doped porous carbon sheets as highly efficient and durable oxygen reduction electrocatalysts", *Angew.Chem.*,53(51), 14235-14239.

# Optimization of molding process for CFRTP having high thermal stability

Toshihira Irisawa[1*], Ryo Hashimoto[1], Masahiro Arai[1], Yasuhiro Tanabe[1]

[1]Graduate School of Engineering, Nagoya University, Nagoya, Japan.
*irisawa.toshihira@material.nagoya-u.ac.jp

## INTRODUCTION

Currently, some commodity thermoplastics are chosen as the matrix polymer of carbon fiber reinforced thermoplastics (CFRTPs) in terms of cost [1]. These CFRTPs, however, have low thermal stability and cannot be used for aircraft applications. Therefore, it is essential to use thermoplastics with much higher thermal stability.

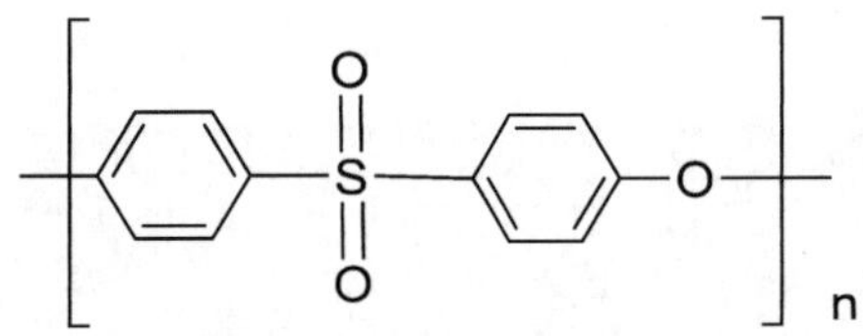

**Fig. 1** Molecular structure of PES.

Polyethersulfone (PES) (Fig. 1) known as amorphous polymers with high $T_g$ of over 200 °C are chosen as the matrix for CFRTPs in this study. The mechanical properties of prototype CFRTPs made with PES have been already reported in our previous study [2]. In contrast, the present study focus on the optimization of molding process for CFRTPs made with PES. The mechanical properties of CFRTPs made under many condition have been evaluated. The interfacial shear strength (IFSS) between CF and PES, and the mechanical properties of PES themselves have been also evaluated. Relationship between measured properties are discussed for optimization of molding process for CFRTPs.

## MATERIALS AND METHODS

### Material

PES with $T_g$ of about 220 °C were used as the matrix polymer, and PAN-based CF (HTS-40-3K, Toho Tenax Co., Ltd.) plain fabrics were used as the reinforcement fibers. Prepregs impregnated with PES (resin Contents of 37 wt%) were supplied by Toho Tenax Co., Ltd. CFRTPs were molded by hot-press: Nine Prepreg sheets cut into 110 mmx 110 mm were cross-ply stacked. The stacked prepregs sheets were hot-pressed in a mold at a target temperature from 300 to 400 °C under 0.5 MPa for 900 s, and under 2.5 MPa for 1200 s at the target temperature in the mold, and then promptly quenched by placing the press mold between two steel plates cooled with water at 25 °C in the presence of 2.5 MPa pressure. CFRTPs with thickness of 1.7 mm and volume fraction was about 60 vol% were obtained.

### Test methods

Mechanical properties of CFRTPs were evaluated by a 3-point bending test at room temperature (R.T). Modulus and strength were determined from the initial slope of the stress-strain curve and the maximum stress attained at the specimen break. The side view of fractured CFRTP was observed by optical microscopy. Mechanical properties of PES films were evaluated by tensile test. Films were made by hot-press at the same condition as molding of CFRTPs. The interfacial shear strength (IFSS) between CF and PES was measured by using fragmentation method at R.T. Since deterioration of PES during the molding of CFRTPs can be evaluated by intrinsic viscosity measurement, intrinsic viscosity of PES was measured by using Ostwald viscometer for PES films having the same thermal history of the CFRTPs preparation. They were dissolved in N, N-dimethylformamide at 30 °C.

## 0RESULTS AND DISCUSSION

The average modulus and strength of CFRTPs molded at each temperature are shown in Fig. 2. The strength of CFRTPs molded at 290 - 330 °C showed almost constant. The strength, however, decreased over 340 °C and then ultimately became about 55 % of molded one at 400 °C. Although the modulus of CFRTPs molded at 350 °C slightly increased as compared with the other, it decreased over 350 °C.

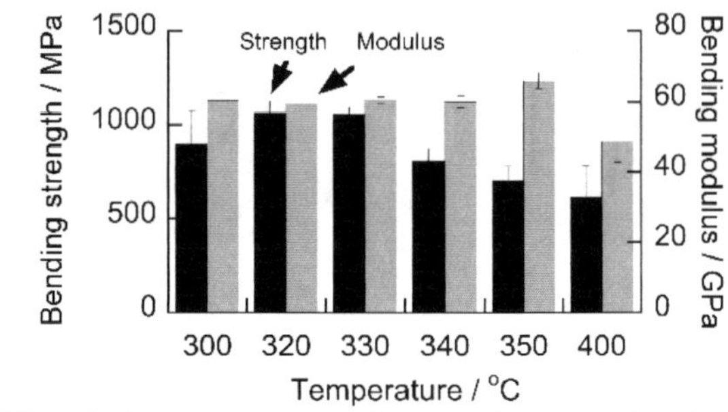


**Fig. 2** Average modulus and strength of CFRTPs molded at each temperature. Error bars mean the standard deviation.

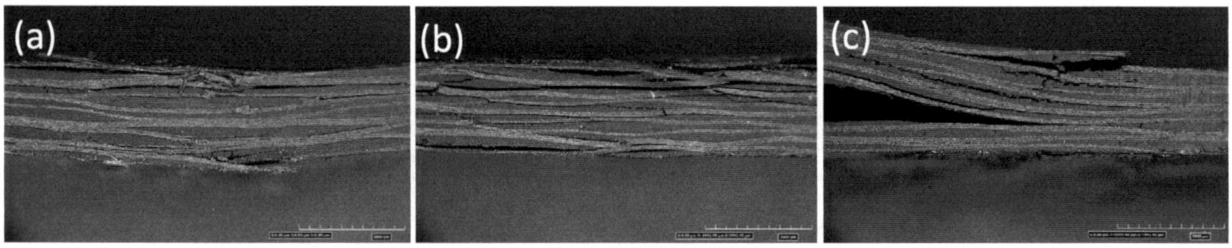


**Fig. 3** the optical microscopic side images of the breaking position after bending tests of CFRTPs molded at (a) 320, (b) 350 and (c) 400 °C.

Fig. 3 shows the optical microscopic side-view images of the breaking position after bending tests. Although there was no vertical crack inside the CFRTP molded at 320 °C, there were significant damages on the top and the bottom sides attribute to fibers fractures and delamination cracks. On the other hand, there were no similar damages in the case of CFRTP molded at 350 and 400 °C but the delamination cracks and matrix fractures could be seen on entire area of the side. Especially, the damage of CFRTP molded at 400 °C becames significant. It is thought that there are two possibilities for decreasing mechanical properties on CFRTPs. One is the decreasing of mechanical properties of matrix polymer, and the other is thermal deterioration of interfacial adhesion, because of increasing of the molding temperature.

**Table 1** Tensile properties, intrinsic viscosity of each PES films and IFSS between CF and each films.

| Temperature [°C] | Modulus [GPa] | Strength [MPa] | Elongation [%] | IFSS [MPa] | [η] |
|---|---|---|---|---|---|
| 300 | 1.56 | 65.8 | 7.33 | 32.7 | - |
| 320 | 1.78 | 66.1 | 6.25 | 34.8 | 0.319 |
| 340 | 1.61 | 71.5 | 7.78 | 39.5 | - |
| 350 | 1.94 | 71.7 | 5.34 | - | 0.283 |
| 400 | 1.93 | 66.3 | 4.50 | - | 0.252 |

The tensile properties, intrinsic viscosity of PES films molded at each temperature and IFSS of CF/PES are shown in Table 1. The elongation and intrinsic viscosity decreased at over 350 °C. Decreasing the intrinsic viscosity means decrease of molecular weight. In this case, the decrease of molecular weight gives no significant influence on the modulus and strength, but on the elongation. It is thought that this fact influence decreasing mechanical properties of CFRTPs.

In summary, deterioration of PES influences the mechanical properties of CFRTP. It is clear that the CFRTP made with PES should mold at under 340 °C.


## ACNOWLEGMENT

This work was supported by CSTI, SIP Program, "Development of High-Rate Production Innovative Aircraft PMC Products and Quality Assurance Technology" (Funding: JST). The materials are supplied by Toho Tenax co., Ltd.

## REFERENCES

1. Zushi, H., Tamura, M., Ohsawa, I., Uzawa, K., Takahashi J. and Yasuda, H. (2006). *J. Jpn. Comp. Mater.*, vol. 32, p.153
2. Irisawa, T., Tanabe, Y., Hashimoto, R. and Arai., M. (2017). *J. Fiber Sci. Technol.*, vol. 73(3), p.61-66

# Oxy-tetracycline Removal via Ozonation Enhanced with Synthesized Magnetic Carbon Nanoparticles

Konrat Kerdnawee[1], Chompoopich Termvidchakorn[1], Karanick Menakanist[1], Napon Opasanon[1], Prakitr Srisuma[1], Nuttawat Suwattanapongtada[1], Noriaki Sano[2], Hajime Tamon[2], Tawatchai Charinpanitkul[1*]

[1] Center of Excellence in Particle Technology, Department of Chemical Engineering, Faculty of Engineering, Chulalongkorn University, Bangkok, Thailand
[2] Department of Chemical Engineering, Graduate School of Engineering, Kyoto University, Kyoto, Japan
*ctawat@chula.ac.th

## INTRODUCTION

An increase in antibiotic usage and exposure within agricultural industries leads to environmental problem because of its over-dosage and abandoned leakage. Therefore, it is necessary to develop an efficient technique to remove such antibiotic. Oxy-tetracycline (OTC) is an antibiotic recognized as a representative contaminant which could be removed by catalytic oxidation. Meanwhile, glycerol is a major by-product of bio-diesel manufacturing process, could be converted to carbon nanoparticles for various applications including pollutant removal. Therefore, a synergetic work focusing on converting abundant glycerol to carbon nanoparticles for tackling such serious problem of antibiotic pollutant was proposed. In this work a facile method of pyrolysis of glycerol and ferrocene was employed for synthesizing magnetic carbon nanoparticles (M-CNPs), which could be collected after usage in removing antibiotic from polluted water. Performance of M-CNPs in OTC removal was experimentally investigated and discussed.

## MATERIALS AND METHODS

### Synthesis of M-CNPs

M-CNPs were synthesized by an experimental set-up consisting of electrical furnaces and quartz tube with size of 11 mm in I.D. and 570 mm in length under $N_2$ gas flow. Glycerol (Qrex, 99.5% purity) and ferrocene (Sigma-Aldrich, 98% purity) with weight ratio of 3:1 were respectively used as precursor of carbon and Fe catalyst for synthesizing M-CNPs. The quartz tube reactor temperature was regulated up to a designated temperature of 800 °C. After a period of designated holding time, the reactor was cooled down to the room temperature and the resultant M-CNPs were collected from the inner wall of the reactor.

### Removal of OTC

0.5 g/m$^3$ of ozone was generated from oxygen by an ozone generator via corona-discharge technique. 10 mg of synthesized M-CNPs, carbon black or graphite powder was dispersed in 20 mL of OTC solution with an initial concentration of 500 mg/L. Then ozone with a flow rate of 20 mL/min was fed into the reactor through a sintered glass filter. Liquid sample withdrawn at designated time was analysed by HPLC (SPD-20A, Shimadzu Corp.) for determining the remaining OTC concentration. Finally, the solid sample was separated from the liquid samples in prior to its characterization. While M-CNPs were separated by external permanent magnetic field, graphitic powder and carbon black were separated using filtration paper.

## RESULTS AND DISCUSSION

Microscopic analyses by TEM (JEOL, JEM1010) suggested that the synthesized product mainly consisted of multi-walled carbon nanotubes with metal nanoparticles. The average diameter of M-CNPs was in a range of 20-60 nm. BET surface of M-CNPs was 22 m$^2$/g. Peak intensity ratio

($I_G$ to $I_D$) of graphitic carbon and disorder carbon structure of M-CNPs which were analysed by Raman spectroscopy (Lamda Vision, MicroRAM-3000L) was higher than unity, suggesting that M-CNPs possessed higher content of graphitic carbon structure or carbon nanotube structure. Under a moderate condition of 800 °C with sufficient amount of the precursors, carbon atoms and clusters could undergo self-assembly process to form carbon nanotubes containing Fe nanoparticles within their carbonaceous structure[1]. In addition, it was found that M-CNPs exerted the AC-magnetic susceptibility, which was expressed in term of real numbers ($X_{r,0}$) of 68 $g^{-1}$. As a result, M-CNPs could respond to external magnetic field.

From our experimental observation, OTC removal was a combination of adsorption and oxidation via ozonation[2] as shown in **Fig.1**. The removal rate of OTC was much improved when M-CNPs were employed as catalyst in comparison to the ozonation alone and adsorption with the absence of ozone. Catalytic performance of the synthesized M-CNPs for OTC removal with ozonation was compared with those of other typical commercial carbonaceous catalysts which were carbon black and graphite powder. Superior catalytic performance of M-CNPs was confirmed by complete removal of OTC within 30 min or two times faster than other conditions using carbon black and graphite powder. These results would be ascribed that embedded Fe in M-CNPs could also promote ozonation[3]. Therefore, it is reasonably implied that M-CNPs could exert capability of adsorption and catalytic decomposition of OTC with the presence of ozone. Moreover, M-CNPs could be easily separated from wastewater by external permanent magnetic bar because of their excellent magnetic property.

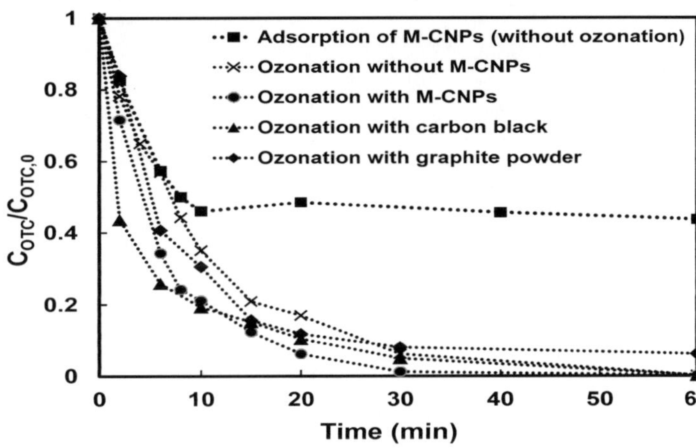


**Fig.1.** OTC removal profiles in various conditions

## CONCLUSIONS

M-CNPs could be synthesized via co-pyrolysis of glycerol with ferrocene. The catalytic ozonation with the presence of M-CNPs could significantly enhance the removal of OTC when compared to adsorption and ozonation alone. In comparison with other carbonaceous catalysts, M-CNPs could exhibit a superior performance in OTC removal. Moreover, M-CNPs could be easily separated by external permanent magnetic bar.

## REFERENCES


1. Charinpanitkul, T., Sano, N., Puengjinda, P., Klanwan, J., Akrapattangkul, N., Tanthapanichakoon, W. (2009). "Naphthalene as an alternative carbon source for pyrolytic synthesis of carbon nanostructures", *J. Anal .Appl. Pyrolysis.*, Vol.86 (2), p.386-390.

2. Wang, Y., Sun, H., Duan, X., Ang, H.M., Tadé, M.O., Wang, S. (2015). "A new magnetic nano zero-valent iron encapsulated in carbon spheres for oxidative degradation of phenol", *Appl. Catal. B: Environ.*, Vol.172, p.73-81.

3. Sanchez-Polo, M., Rivera-Utrilla, J. (2003). "Effect of the ozone–carbon reaction on the catalytic activity of activated carbon during the degradation of 1,3,6-naphthalenetrisulphonic acid with ozone", *CARBON.*, Vol.41, p.303-207.

# Performance of Pd/graphene oxide catalysts: influence of reduction degree

Cristina Ruiz-García[1*], Yu Lei[2], Francisco Heras[1], Ana Laura Elías[3], Mauricio Terrones[2,3], Miguel A. Gilarranz[1]

[1] Sección de Ingeniería Química, Universidad Autónoma de Madrid, Spain.
[2]Department of Materials Science and Engineering, The Pennsylvania State University, State College, USA.
[3]Department of Physics and Center for 2-Dimensional and Layered Materials, The Pennsylvania State University, State College, USA.
*cristina.ruiz@uam.es

## INTRODUCTION

Graphene oxide (GO), obtained by oxidation and exfoliation of graphite, is the most common way to obtain bulk amounts of graphene. GO has properties that make it attractive as a catalyst or catalyst support. The presence of abundant oxygen groups in GO can help to disperse metals on its surface. Likewise, reduced GO (rGO) exhibits differential properties that can be of interest to modulate the interaction with such metallic phases. The chemical reduction of GO with a variety of reducing agents, such as hydrazine and sodium borohydride, has been reported. Besides reducing GO effectively, hydrazine reduction can lead to the insertion of different N groups such as amines, hydrazones, among others into rGO[1)]. The heteroatoms in rGO can play an important role in the metal phase dispersion and stabilization, and are also important for catalytic activity[2)]. The aim of this work is to obtain a set of rGO-Pd materials based on GO with different reduction degrees and nitrogen functionalities, following a one-pot simple process. The materials are tested as catalysts in the aqueous phase hydrodechlorination (HDC) of 4-chlorophenol (4-CPh) in order to establish relationships between their chemical surface, structural characteristics and catalytic performance.

## MATERIALS AND METHODS

GO was prepared by oxidation and exfoliation of graphite bulk powder using the procedure reported before by Cruz-Silva et al.[3)]. To synthesize rGO-Pd materials, GO was firstly dispersed in ethanol using an ultrasonic bath, and then a solution of $Pd(AcO)_2$ was added to the GO suspension to achieve 1.5 *wt.* % Pd (GO basis). This suspension was used to obtain different rGO-Pd materials: rGOEt/Pd was obtained by reduction with the ethanol contained in the dispersion at 25°C; rGONH25/Pd and rGONH60/Pd were obtained through reduction with $N_2H_4$ at 25 °C and 60 °C, respectively; and rGOBH/Pd was obtained by reduction with $NaBH_4$. The materials were filtered, washed with DI water and dried. Characterization of rGO-Pd materials was carried out by X-ray photoelectron spectroscopy (XPS), nitrogen adsorption-desorption, bulk elemental chemical analyses, thermogravimetric analysis (TGA), transmission electron microscopy (TEM) and X-ray diffraction (XRD). The rGO-Pd materials were tested as catalyst in the HDC of 4-CPh. The reaction experiments were carried out in a batch isotherm reactor at 30 and 70°C stirred reactor using 150 mL of a 0.1 g/L 4-CPh aqueous solution, 3.0 mg/L of Pd, 50 mL/min of hydrogen feed and 800 rpm of stirring. Activities were calculated from pseudo-first order rate constants for the disappearance of 4-CPh.

## RESULTS AND DISCUSSION

Table 1 shows the characterization of the rGO-Pd materials prepared. rGO-Pd materials were successfully prepared with the proposed one-pot procedure. XRD data clearly shows the GO reduction after the treatment with $N_2H_4$ and $NaBH_4$, exhibiting the d(002) close to 0.34 nm. The reduction of the GO is also evidenced by the decreasing in oxygen content observed in XPS line scans. In the case of the reduction with $N_2H_4$, nitrogen was also incorporated in the rGO, mostly as pyrrolic (N-5), imine, amide or amine (N-X) and quaternary nitrogen (N-q), although pyridinic and nitrogen oxide signals were also detected. A slightly lower oxidation temperature was observed by TGA for nitrogen-doped rGOs. Surface area values decreased to very low values after reduction, showing only a meso-porosity contribution. TEM images depicted well dispersed Pd nanoparticles, evidencing reduction of the metal precursor. Interestingly, the Pd nanoparticles were generated in a very narrow size range (5.2±2.6 nm) when $NaBH_4$, was used as reducing agent, even though no capping agent was added, indicating that the rGO influences nucleation and limits the growth of the metallic phase.

**Table 1.** Characterization of rGO-Pd materials

| **Sample** | $S_{BET}$ ($m^2/g$) | $S_{EXT}$ ($m^2/g$) | **O** (%,w) | **N** (%,w) | **C/O** | $d_{002}$ (nm) | **Pd size** (nm) |
|---|---|---|---|---|---|---|---|
| rGOEt/Pd | 267 | 127 | 32.4 | 0.7 | 1.8 | 0.90 | 19.0±14.4 |
| rGONH25/Pd | 16 | 17 | 21.1 | 3.4 | 3.2 | 0.37 | 64.8±58.1 |
| rGONH60/Pd | 16 | 22 | 10.2 | 2.7 | 8.4 | 0.37 | 31.6±23.6 |
| rGOBH/Pd | 88 | 76 | 14.3 | 0.3 | 5.6 | 0.36 | 5.2±2.6 |

The rGO-Pd materials were tested in the HDC of 4-CPh at 30 and 70°C. All of them showed activity in HDC, but limited hydrogenation capacity, with phenol as the main product. rGOEt/Pd was the most active at 30°C (11 mmol/gmin), providing with complete disappearance of 4-CPh in 60 min, however its activity only increased to 13 mmol/gmin when temperature was increased to 70°C. The catalysts prepared by reduction with $N_2H_4$ and $NaBH_4$ showed in general lower activity than the one reduced with ethanol at 30°C (1-11 mmol/gmin). Interstingly, their activity was boosted to 5-36 mmol/gmin at 70 °C. In the case of GOBH/Pd, the improved performance can be attributed only to reduction. However, the activity of GONH60/Pd increased remarkably at 70 °C (36 mmol/gmin) and was the most active, showing a correlation with the oxidation degree and also the role of N doping. The activity value obtained for GONH60/Pd exceeded all expectations, considering the low surface area of the catalyst and also the large Pd nanoparticle size.

## CONCLUSIONS

A simple procedure for the one-pot synthesis of rGO-Pd materials has been proved. Small and controlled size Pd NPs were obtained without a capping agent when sodium borohydride was used as a reducing agent. rGOEt/Pd was the most active material at 30°C in the HDC of 4-CPh, whereas the most reduced rGO led to catalysts whose activity was boosted at high reaction temperature, particularly in the case of nitrogen doped rGO.


## ACKNOLEDGEMENT

The authors are grateful to MINECO for support (CTQ2012-32821, CTQ2015-65491-R), and C. Ruiz-García also for PhD grant (BES-2013-066085; EEBB-I-16-11722).


## REFERENCES


1. Pei, S., and Cheng, H.-M. (2012). "The reduction of graphene oxide". Carbon, Vol. 50 (9), p.3210-3228.
2. Huang, C., Lia, C. and Shi, G. (2012). "Graphene based catalysts" Energy Environ. Sci., Vol. 5 (10), p.8848-8868
3. Cruz-Silva, R., Morelos-Gomez, A., Kim, H.-I., Jang, H.-K., Tristan, F., Vega-Diaz, S., Rajukumar, L.P., Elías, A.L., Perea-Lopez, N., Suhr, J., Endo, M. and Terrones, M. (2014). "Super-

stretchable Graphene Oxide Macroscopic Fibers with Outstanding Knotability Fabricated by Dry Film Scrolling". ACS Nano, 8(6), p.5959-5967.

# Polyaniline-based carbon nanospheres as anode materials for sodium-ion batteries

Chunli Zhou, Huaihe Song*, Xiaohong Chen, Mengqiu Jia

State Key Laboratory of Chemical Resource Engineering, Beijing Key Laboratory of Electrochemical Process and Technology for Materials, Beijing University of Chemical Technology, Beijing, 100029, P.R. China
*songhh@mail.buct.edu.cn

## INTRODUCTION

Larger ionic diameter of $Na^+$ relative to $Li^+$ is the main bottleneck in the development of sodium-ion batteries, so many researchers devoted to designing and preparing great electrode materials[1)]. Hard carbon materials are studied widely due to the disordered structure and larger interlayer distance[2)]. However, the rate performance of hard materials in previous reports is desirable to be enhanced. Heteroatom-doped carbon materials have been widely investigated for abundant extrinsic defects and increased interlayer distance[3)]. Based on this background, we have prepared polyaniline-based carbon nanospheres (PCSs) by simple pyrolysis treatment of polyaniline nanospheres. PCSs show the good capacitance property and excellent rate performance which are superior to that of other carbon electrode materials in literature. Through structural characterization and detailed analysis, we consider that the excellent electrochemical performance of PCSs is due to high N/O-codoped level, disordered structure, large interlayer distance and distinct superiorities of spherical structure.

## MATERIALS AND METHODS

Polyaniline nanospheres were synthesized as our previous work via controlling the volume of aniline monomer (0.5 mL) [4)]. Afterwards, polyaniline nanospheres were heated at 200 □ for 2 h in oxygen atmosphere to obtain stabilizing product. And then oxidized nanospheres were carbonized at 800□ for 2 h in argon atmosphere to get the target hard carbon nanospheres (labeled as PCSs).

## RESULTS AND DISCUSSION

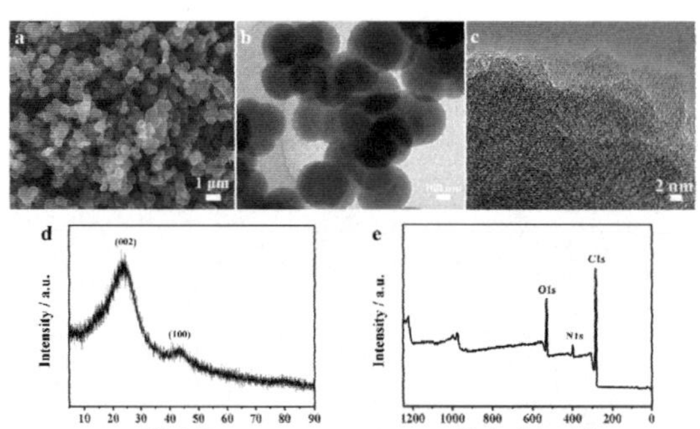


**Fig. 1.** (a) SEM image of PCSs, (b) TEM image of PCSs, (c) HRTEM image of PCSs, (d) XRD pattern of PCSs, (e) XPS spectra of PCSs, respectively.

SEM image (**Fig. 1**a) shows that PCSs were prepared successfully via simple calcination treatment of polyaniline nanospheres, and TEM image (**Fig. 1**b) proves the solid structure of PCSs. HRTEM image taken at the edge of PCSs distinctly exihibits the amorphous structure (**Fig. 1**c). XRD pattern

(**Fig. 1**d) has two dispersive diffraction peaks, ascribing to the (002) and (100) crystal planes of carbon, respectively. This result can certify the low crystallinity and disordered amorphous structure of PCSs and the distance between adjacent carbon platelets is calculated via Bragg equation to be 0.362 nm. This value is higher than that of graphite, suggesting the excellent electrochemical performance of PCSs when used as anode material for sodium-ion batteries. The chemical state of the surface of PCSs is further analyzed by XPS. The survey spectra (**Fig. 1**e) shows the N/O-codoped characteristic of PCSs and heteroatom content reaches up to 21.15%, which can bring more defects and active sites.

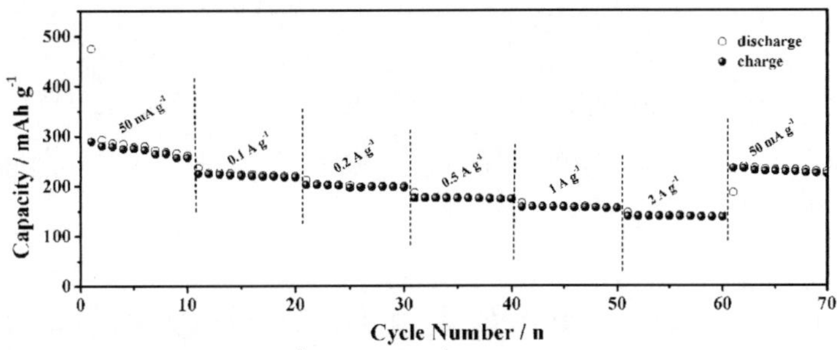


**Fig. 2.** rate performance of PCSs at 50 mA g$^{-1}$ to 2 A g$^{-1}$.

The specific capacity of PCSs is 289 mAh g$^{-1}$ at 50 mA g$^{-1}$, revealing the excellent capacity property. Reversible capacity still maintains 139 mAh g$^{-1}$ at 2 A g$^{-1}$ and could recover to 224 mAh g$^{-1}$ when the current density returns to 50 mA g$^{-1}$, illustrating that PCSs possess excellent rate performance. The great electrochemical performance of PCSs could be attributed to spherical structure, disorder construction, large interlayer distance and high heteroatom-doped content.

## ACKNOWLEDGMENTS

This work was supported by the National Natural Science Foundation of China ( U1610252), and Foundation of Excellent Doctoral Dissertation of Beijing City (YB20121001001).

## CONCLUSIONS

Disordered structure, large interlayer distance and high heteroatom-doped content endow PCSs with high reversible capacity (289 mAh g$^{-1}$ at 50 mA g$^{-1}$) and excellent rate performance (139 mAh g$^{-1}$ at 2 A g$^{-1}$). In a word, synthesis of hard carbon materials with special structure and high heteroatom-doped content is an effective method to enhance electrochemical capacity of NIBs.

## REFERENCES


1. Cao Y, Xiao L, Sushko ML, Wang W, Schwenzer B, Xiao J, et al. (2012). "Sodium ion insertion in hollow carbon nanowires for battery applications", *Nano lett.*, Vol.12(7), p.3783-7.
2. Irisarri E, Ponrouch A, Palacin MR. (2015). "Review—Hard Carbon Negative Electrode Materials for Sodium-Ion Batteries", *J. Electrochem. Soc.*, Vol.162(14), p.A2476-A82.
3. Wang M, Yang Z, Li W, Gu L, Yu Y. (2016). "Superior Sodium Storage in 3D Interconnected Nitrogen and Oxygen Dual-Doped Carbon Network", *Small*, Vol.12(19), p.2559-66.
4. Zhou C, Chen X, Liu H, Zhou J, Ma Z, Jia M, Song H. (2017). "Heteroatom-doped multilocular carbon nanospheres with high surface utilization and excellent rate capability as electrode material for supercapacitors", *Electrochim. Acta,* Vol.236, p.53-60.

# Potential of Carbon Gels in Hydrogen Storage Systems

Krisztina László[1*], Orsolya Czakkel[2#], Balázs Nagy[1], Emanuel Bahn[2,3], Peter Fouquet[2], Silvia Villar-Rodil[4], Juan M. D. Tascón[4],

[1]Department of Physical Chemistry and Materials, Budapest University of Technology and Economics, Budapest, Hungary
[2]Institut Laue-Langevin, Grenoble, France
[3]Cavendish Laboratory, Cambridge UK
[4]Instituto Nacional del Carbón, INCAR-CSIC, Oviedo, Spain
*klaszlo@mail.bme.hu; #czakkelo@ill.fr;

## INTRODUCTION

Hydrogen is considered to be one of the most promising "green" fuels (combustion yields water vapour), as its energy content (142 MJ/kg) exceeds that of petroleum by a factor of three. However, safe efficient storage and use of $H_2$ still raises several technological challenges. and efficient hydrogen storage with high capacities still remains a key technological bottleneck for practical applications[1)]. Hydrogen storage by sorption of molecular $H_2$ in solid-state porous materials is offering attractive storage possibility owing to its fast kinetics, good cyclability and high adsorption capacity[1)].

In this work, we focus on the hydrogen adsorption properties of carbon gels prepared from resorcinol-formaldehyde resins. To compare the effect of the pore characteristics and the surface chemistry samples were prepared with different drying methods and post carbonization treatments. The carbon gels were characterized by low temperature $N_2$ adsorption/desorption, $CO_2$ adsorption and X-ray photoelectron spectroscopy (XPS). $H_2$ uptake measurements were performed at different temperatures. The mobility of the adsorbed $H_2$ on the surface was studied by neutron spin-echo (NSE) spectroscopy.

## MATERIALS AND METHODS

### Material

Carbon aerogel samples were prepared according to the method described elsewhere[2)]. To obtain different pore morphologies prior to carbonisation at 1173 K for 1 h in a 25 ml/min flow of high purity nitrogen (99.996%, Linde), the wet gels were dried either supercritically (yielding the sample CA) according to the method described by Czakkel et al.[3)], or in ambient conditions (yielding the sample CX). To modify the surface chemistry of CA a portion of the sample was treated by cc. $HNO_3$ for 3 h at 20°C (sample CAox). Acid residue was removed by washing with distilled water in a Soxhlet extractor until the pH became neutral.

### Methods

*Nitrogen adsorption/desorption* isotherms were measured at 77 K with a Nova 2000e (Quantachrome) computer controlled surface area and pore size analyser. The apparent surface area ($S_{BET}$), total pore volume ($V_{TOT}$), the micropore volume ($W_{0-DR}$), and the pore size distribution in the micro- and mesopore regions were derived as described earlier[4)]. *Carbon dioxide adsorption/desorption* isotherms were measured at 273 K with an Autosorb-1 surface area and pore size analyser (from Quantachrome). The volume of the ultramicropores was determined from the Dubinin-Raduskevich plot in the $p < 0.03$ bar region. Transformation of the primary adsorption data was performed by the Quantachrome software ASiQwin version 3.0.

*$H_2$ adsorption/desorption* measurements were performed by a Sieverts (volumetric) sorption analyser from Hiden Isochema in the 60-90 K temperature range, up to 20 bar pressure. Prior to each measurement, He adsorption/desorption isotherms were measured as well in the same pressure and temperature range, which served as correction.

*Neutron spin-echo spectroscopy (NSE)* measurements were made on the spectrometer IN11, as described earlier[4)], at the Institut Laue-Langevin (ILL), Grenoble, France. The temperature was controlled by a standard orange cryofurnace in the 20-80 K range. During NSE measurements a Sieverts (volumetric) sorption analyser (HIDEN Isochema) was used. The adsorption of $H_2$ was carried out at 20 K; the injected amount was calculated such that the $H_2$ molecules cover half of the available surface area of the sample. Calculation of the $H_2$ amount was based on the $S_{BET}$ of the samples and the area/molecule occupied by the $H_2$ molecule in the adsorbed state (15.5 A□$^{2}$)[5)].

## RESULTS AND DISCUSSION

High pressure $H_2$ uptake measurements at 77 K showed outstandingly high $H_2$ capacities for all the three samples (Fig. 1). While the equilibrium $H_2$ adsorption characteristics depend mainly on the pore properties of the adsorbent, diffusion of the adsorbed $H_2$ molecules is affected by both the surface chemistry and the porosity of the carbon. The isosteric heat of adsorption extrapolated to zero coverage is higher (and closer to the theoretical optimum value) in the present samples than is usually observed in similar systems, revealing the potential of these materials in future $H_2$ storage systems.

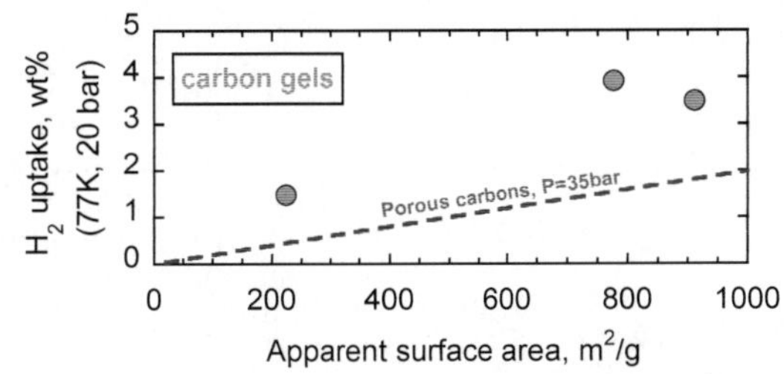


**Figure 4 $H_2$ uptake at 77 K and 20 bar as a function of $S_{BET}$. Dashed line shows the hydrogen uptake at 35 bar usually observed in porous carbons (1wt% $H_2$/500 $m^2$*$g^{-1}$ $N_2$ BET surface**

## CONCLUSIONS

Our experiments proved that NSE is a very powerful tool to study the surface diffusivity of adsorbed $H_2$, even at low relative coverage. The results reveal that the surface chemistry of the carbon has substantial impact on the surface diffusion properties of the hydrogen, while by changing the pore characteristics of the carbon the hydrogen desorption can be shifted towards significantly higher temperatures.

## ACKNOWLEDGEMENT


The authors are grateful to the Institut Laue-Langevin and ISIS Neutron and Muon Source facilities access to the instruments. This research was supported by the Hungarian Scientific Research Fund K109558. The work is part of the EU project FP7-PEOPLE-2010-IRSES-269267 and the Nanoporous and Nanostructured Materials for Medical Applications (Nanomed: H2020-MSCA-RISE-2016, project:734641).


## REFERENCES


1. Xia, Y., Yang, Z., Zhu, Y. (2013). "Porous carbon-based materials for hydrogen storage: advancement and challenges", *J. Mater. Chem. A,* Vol.1, p.9365-9381

2. Czakkel, O., Marthi, K., Geissler, E., László, K. (2005). "Influence of drying on the morphology of resorcinol–formaldehyde-based carbon gels", *Micropor. Mesopor. Mater.,* Vol.86(1-3), p.124-133

3. Czakkel, O., Székely, E., Koczka, B., Geissler, E., László, K. (2012) "Drying of resorcinol–formaldehyde gels with $CO_2$ medium", *Micropor. Mesopor. Mater.* Vol.148(1), p.34-42

4. Bahn, E., Czakkel, O., Nagy, B., László, K., Villar-Rodil, S., Tascón, J. M. D., Demmel, F., Telling, M.T.F., Fouquet, P. (2016). "Diffusion of molecular hydrogen in carbon aerogel" *Carbon,* Vol.98, p572-581

5. Nielsen, M., McTague, J.P., Ellenson, W. (1977). "Adsorbed layers of $D_2$, $H_2$, $O_2$, and $^3He$ on graphite studied by neutron scattering" *J. Phys. Colloques*, Vol.38(C4), 10-18
6. Panella, B, Hirscher, M., Roth, S. (2005). "Hydrogen adsorption in different carbon nanostructures" *Carbon*, Vol.43(10), p2209-2214

# PREPARATION AND APPLICATION OF FLEXIBLE AND SELF-STANDING NON-WOVEN CARBON NANOFIBERS

**YANG, X**[1], Liu, J[1], Wang, X[1] , Liang, J[1], Xue, Y[1]

18518572325@163.com

[1]*State Key Laboratory of Organic-Inorganic Composites, Beijing University of Chemical Technology, Chao-Yang District, Beijing 100029, China Flinders University,*

## INTRODUCTION

The preparation of electrospun nanofiber mats is the first step to obtain high strength carbon nanofiber mats. The shape and mechanical properties of electrospun nanofiber mats play a decisive role in the final electrospinning of ultrafine carbon nanofiber mats. At the same time, pre-oxidation is also one of the important steps, which the structural change determines the structure and properties of the final carbon fiber to a great extent.As in previous studies, the stabilization process need a long time, this study found that by shortening the stabilization time, controlling the oxygen content during the stabilization process, the tensile strength of carbon nanofiber mats can be enhanced.

## MATERIALS AND METHODS

The PAN powder was dried and dissolved in the mixed solution between DMF and THF to prepare a 10wt% or 11wt% solution. The electrospun sheets were peeled from the aluminium foil and the stabilization was conducted from 240 °C to 280 °C with a constant flow of air through the chamber. Refer to the previous thermal stability process, The heating rate was 1 °C/min and finally stayed at 280 °C for 2 h to complete the stabilization. In addition, we consider shortening the stabilization time, controlling the oxygen content, and improving the strength of carbon nanofiber mats. The heat stabilization time was 20min, 17min, 15min, 40min, respectively. During the stabilization, the highest tensile force of 196.64 MPa was imposed. The stabilized nanofibre sheets were held at constant length and a form of ladder-type heating and a constant flow of high-purity nitrogen were applied in this step. 5 °C/min from the room temperature to 300 °C and stayed at 300 °C for 2 hours, then 5 °C/min from300 to 800°C and stayed at 800°C for 1hour, 4°C/min from 800 to 1000°C and stayed at 1000°C for 1hour, 1.5°C/min from 1000 to 1200°C and stayed at the last temperature for 2 min. After pyrolysis, the temperature was allowed to cool down to room temperature at the same speed before the sample was exposed to ambient air.

## RESULTS AND DISCUSSION

### FT-IR

Figure 1 showsthe absorbance peaks centred at 2240 and1590 cm, corresponding to the stretching motions/vibrations of C≡N bonds and C=N bonds, are two important peaks to characterize the cyclization occurring during stabilization.

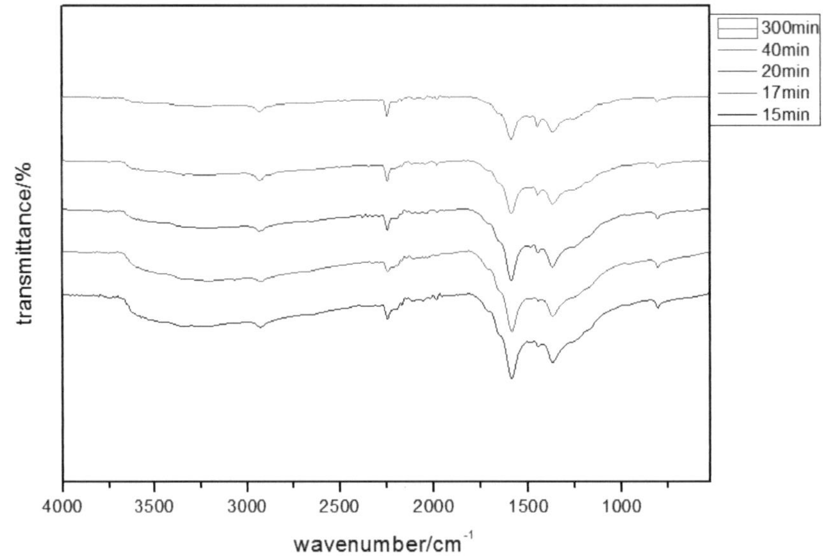


Fig. 1.ATR-FTIR spectra of nanofibers stabilized with different residence times

**SEM**

Fig. 2a and b shows the images of stabilized nanofibers which were stabilized 240~280 °C for 40min. Owing to shorten the stabilization time during stabilization, even though fibres shrank due to the cyclization reaction and the average diameter of the fibres was further reduced to about 280 nm.

1.

(b)

Fig. 2. SEM microphotographs of (a and b) the nanofibres carbonized at 1200 °C for 2 min.

**CONCLUSIONS**

(1) ideal PAN nanofibers were obtained.

(2) In addition, under our optimized temperature profile, flexible and self-standing carbon nanofibers were prepared with significantly reduced stabilization time. The longer the thermal stability time, the higher the oxygen content, the higher the micropore content of the carbon nanofiber felt and the lower the mechanical properties. The shorter the time of thermal stability, the smaller the degree of cyclization, the lower the yield of carbon and the deviation of mechanical properties.

(3) The results show that when the optimized time and temperature during the stabilization process are 274°C and 40minthe responding strength of the carbon fiber mats can obtain the maximum value196.64MPa.

**REFERENCES**

1.Molnar K, Vas L M, Czigany T. Determination of tensile strength of electrospun single nanofibers through modeling tensile behavior of the nanofibrous mat[J]. Composites Part B Engineering, 2012, 43(1):15-21.

# Preparation and Investigation of SWCNT Coatings Suitable for Biosensor Applications

Jurgis Barkauskas*, Ieva Paklonskaitė, Justina Gaidukevič, Arūnas Ramanavičius, Lina Mikoliūnaitė, Povilas Genys, Jūratė Jolanta Petronienė, Inga Morkvėnaitė–Vilkončienė

[1]Vilnius University, Faculty of Chemistry and Geosciences, Vilnius, Lithuania
*jurgis.barkauskas@chf.vu.lt

## INTRODUCTION

Carbon nanomaterials (CNMs) are widely used in biosensors as an excellent transducer component. They have diverse and robust electronic properties as well as tuneable properties of surface chemistry. The use of CNMs for biosensors in the form of quantum dots (Qds), carbon nanotubes (CNTs), graphene, graphene oxide (GO), and the like is reported so far[1)]. The electroactivity and increased conductivity of CNMs are readily exploited in biosensors. Therefore, it is very important to understand more precisely the interaction between CNM surface and biomolecules, which are applied in biosensor design.

## MATERIALS AND METHODS

*Preparation of SWCNT, SWCNT-GO and SWCNT-rGO coatings.* SWCNT coatings were prepared on the polycarbonate membrane filter (Merck Millipore Ltd., pore diameter 400 nm). The preparation procedure is described in more detail elsewhere[2)]. Attachment of GO flakes on the top of SWCNT coating was carried out from the aqueous GO solutions using the filtering procedure (SWCNT-GO coatings). SWCNT-GO coatings were annealed up to 160 $^{o}$C (SWCNT-rGO coatings).

*Immobilization of glucose oxidase (GOX).* The surface of coatings was kept in a closed vessel over a 25% solution of glutaraldehyde for 10 min. During this step, the amino groups of the enzyme were interacting with aldehyde groups attached to modified SWCNT surface, which lead to covalent immobilization of the enzyme.

*Characterisation.* Prepared coatings were characterized by means of conductivity measurements, EIS spectroscopy, contact angle measurements (CAM), SEM and Scanning electrochemical microscopy (SECM).

## RESULTS AND DISCUSSION

Produced coatings were investigated in order to identify their structure, electrical properties, surface chemistry and biocompatibility. Electrical conductivity measurements have shown that SWCNTs in contact with a metal (Ag) behave as Shottky barrier devices. Measurement across the SWCNT coating revealed an obvious decrease in resistivity at higher temperatures. The results lead to the conclusion about the high structural anisotropy of the SWCNT coatings.

Experimental points in a Nyquist plot were fitted to the equivalent electrical circuit for each coating. Two types of circuits were appropriate for fitting (**Table 1**). EIS measurements revealed that reduced GO sheets facilitate the electron transport along the annealed SWCNT-rGO coating, and these coatings should be especially suitable for biosensors.

The accepted CAM technique is based on the theory proposed by van Oss and co-workers. This method is referred as an acid-base approach[3)]. We have evaluated applicability of three different solvents of different polarity (water, glycerol and 1-bromnaphtalene). CAM measurements revealed that surface free energy and its components undergo a characteristic change during the annealing procedure. In conclusion, SWCNT coating should be considered as a polar surface with prevailing basic properties. At higher temperatures (180 °C) the acid-base interaction decreases, and non-polar component remains predominant.

**Table 1.** Parameters of EIS fitting.

| | **SWCNT** | **SWCNT-GO** | **SWCNT-rGO** |
|---|---|---|---|
| **Equivalent circuit** | $R_s$–[$R_{p1}$//$C_1$] | $R_s$–[$R_{p1}$//$C_1$] | $R_s$–[$R_{p1}$//$C_1$]–[$R_{p2}$//$C_2$] |
| **$R_S$ (Ω)** | $1.650 \cdot 10^2$ | $1.172 \cdot 10^2$ | $3.00 \cdot 10^2$ |
| **$R_{P1}$ (Ω)** | $1.385 \cdot 10^4$ | $7.966 \cdot 10^6$ | $7.65 \cdot 10^6$ |
| **$R_{P2}$ (Ω)** | - | - | $1.56 \cdot 10^6$ |
| **$C_1$ (F)** | $3.378 \cdot 10^{-10}$ | $7.255 \cdot 10^{-11}$ | $8.79 \cdot 10^{-11}$ |
| **$C_2$ (F)** | - | - | $8.58 \cdot 10^{-11}$ |
| **Correlation coefficient** | 0.941 | 0.983 | 0.985 |

SEM images of SWCNT, SWCNT-GO and SWCNT-rGO coatings are shown in Error! Reference source not found.**1**. The SEM view from the opposite side (**Fig. 1a**) confirms the hypothesis about the protrusion of SWCNT filaments across the pores of the polycarbonate filter. Using SEM it is possible to distinguish the tangled SWCNT agglomerations in the coating morphology. Single GO sheets are observed in SWCNT-GO coatings before annealing. After annealing reduced GO sheets are attached more closely to the SWCNT substrate.

**Fig. 1.** SEM images of coatings. SWCNT coating from the opposite side (polycarbonate) (a); SWCNT (b); SWCNT-GO (c); SWCNT-rGO (d).

Experiments using SECM were carried out with SWCNT, SWCNT-GO and SWCNT-rGO using the pristine coatings, as well as coatings immobilized with GOx, and the latter in the presence of glucose. The results obtained suggest that in the case of immobilized SWCNT-rGO coating a direct electron transfer between GOx and SWCNT is observed.

## CONCLUSIONS

The assumption was made that SWCNT-rGO coatings are prospective for the design of efficient and reliable glucose biosensors. Our future work is planned to prepare and test the implantable glucose amperometric biosensors using the SWCNT-rGO coatings, which were preliminary investigated in this research.


## ACKNOWLEDGEMENT

This research was funded by a grant (No. SEN-21/2015) from the Research Council of Lithuania.


## REFERENCES


1. Baptista, F.R., Belhout, S.A., Giordani, S., Quinn, S.J. (2015) "Recent developments in carbon nanomaterial sensors", *Chem. Soc. Rev.* Vol.44, p.4433-4453.
2. Barkauskas, J. (2010) "pH-dependent water penetration through CNT sub-layers arranged on the polycarbonate membrane filters", *Carbon* Vol.48, p.1858-1865.
3. Hollander, A. (1995) „On the selection of test liquids for the evaluation of acid-base properties of solid-surfaces by contact-angle goniometry", *J. Coll. Interface Sci.* Vol.169, p.493-496.

# Probabilistically based defect analysis and structure-property-relations of CF

Wolz, D. S.[1,2], Thieme, M.[1,2], Kirsten, M.[2,5], Löffler, M.[3], Mühle, U.[4], Rellinghaus, B.[3], Jäger, H.[1,2], Gude, M.[1,2], Zschech, E.[3], Cherif, C.[2,5], Böhm, R.[1,2]

[1] Institute of Lightweight Engineering and Polymer Technology, Technische Universität Dresden, Dresden, Germany

[2] Research Center for Carbon Fibers Saxony, Technische Universität Dresden,

Dresden, Germany

[3] Dresden Center for Nanoanalysis, Technische Universität Dresden, Dresden, Germany

[4] Institute of Materials Science, Technische Universität Dresden, Dresden, Germany

[5] Institute of Textile Machinery and High Performance Material Technology, Technische Universität Dresden, Dresden, Germany

*daniel_sebastian.wolz@tu-dresden.de

## INTRODUCTION

For novel carbon fibre (CF) applications like porous membranes for desalination and fuel cells or low-cost carbon fibres with high-tensile strengths for carbon concrete composites, the fibre properties need to be tailored which requires the structure-property-relations to be analysed. Despite tremendous research effort in recent years, this relationship between CF structure and mechanical properties is not fully understood yet. Therefore, tailoring the CF properties is still very limited. It is well known that each content of a CF structure (crystalline and amorphous contents) and defects on different length scales differently contribute to the CF properties.

## MATERIALS AND METHODS

### *Material*

The investigations have been performed on commercial intermediate-strength-CF from Toho Tenax with an epoxy sizing of a single production batch. The CF consisted of 12,0000 filaments with a linear density of 800 tex.

### *Methods*

*Filament tensile tests*. The filament tensile tests were conducted with FAVIMAT+ tensile test system from Textechno. Applied gauge length was set to 20mm with a pretension force of 0.5 cN/tex and a tension velocity of 10 mm/min. Clamp pressure was adjusted to 5 bar. 328 Filaments has been tested for statistical analysis.

*Optical Analysis.* The CF-filaments were examined by transmission electronic microscopy (TEM) and scanning electron microscopy (SEM). The examined pieces for SEM were prepared from a unidirectional reinforced epoxy composite with carbon fibres from the same filament batch. Samples for TEM and SEM have been extracted with focused ion beam method.

## RESULTS AND DISCUSSION

The performed tensile tests on filament level lead to a large scatter of the mechanical properties and the linear density, see Fig. 1 left. For same linear density different mechanical properties are measured. Assuming homogenous structures within the CF-filament and constant filament diameters, different mechanical properties are a result of different void contents. Images across the filament diameter and along the filament axis show different distributions for void appearances both for individual CF-filaments and overall for the CF-roving at microscale, see Fig. 1 mid.

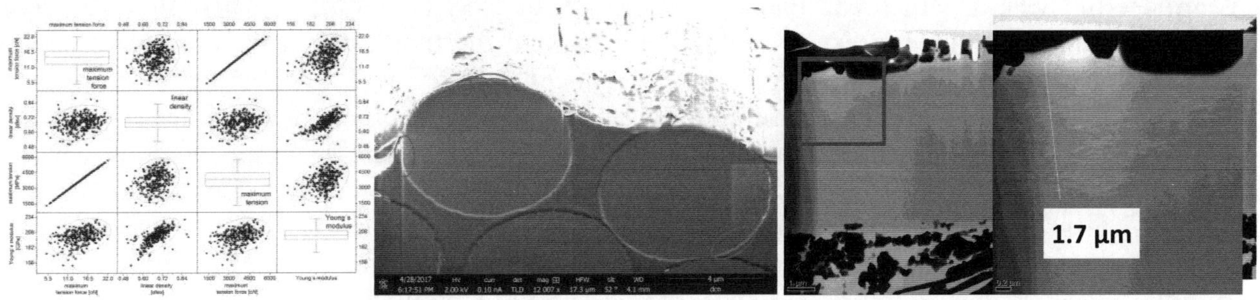


**Figure 1: Correlation matrix for linear density, axial force, tensile strength and Young´s modulus (left); voids across the carbon filament diameter (mid) as well as defect phenomena in carbon fibre filaments along the filament axis (right)**

Besides the differences in defect distribution at the microscale, different flaw phenomena additionally appear on the sub-microscale. Along the filament axis, local flaw clusters appear next to intact zones, see Fig. 1. right. These flaw clusters have the dimension of approx. half of the filament diameter and are one likely origin for bigger local defects which appear on the microscale, see Fig.mid and right. Further correlations for defect phenomena (voids and defects) will be shown within the oral presentation.

## CONCLUSIONS

To build a basis for tailoring the properties of carbon fibres, correlations between the defect phenomena (voids and flaws) and the mechanical properties of the fibre (structure-property-relations) need a deeper investigation. Future studies should concentrate on the probabilistic phenomena of flaw clustering and void formation in the sheath regions and on the contribution of these defects to filament failure. This implies the need of nanoanalysis technology like nano-computer-tomography with resolutions down to single nanometers in order to directly correlate failure and defect phenomena.

# Processing and Characterization of Long Fiber Thermoplastic Technology-based Carbon Fiber/ABS Composites

Donghwan Cho[*], Daekyun Hwang

Department of Polymer Science and Engineering, Kumoh National Institute of Technology, Gumi 39177, Korea
*dcho@kumoh.ac.kr

## INTRODUCTION

Recently carbon fiber reinforced plastics (CFRP) with thermoplastic matrices have been increasingly attracted due to their faster processing time, extrusion/injection processibility, and recyclability, compared to CFRP with thermosetting matrices, particularly in automobile industries. However, use of long fiber reinforcement in CFRP with a thermoplastic matrix needs great scientific and technological challenges since thermoplastic resins exhibit much higher melt viscosity than thermosetting resins, which can make impregnation of the fiber into the matrix more difficult and complicated. Therefore, processing methods for making thermoplastic CFRP are limited. Extrusion is a representative processing method to produce CFRP with chopped carbon fibers and a thermoplastic matrix through compounding, extruding, cooling, and pelletizing processes. Unfortunately, carbon fibers fed in the extruder experiences severe shearing forces in between the screw and the barrel of the extruder, resulting in fiber breakage and shortening. As a result, the mechanical properties of the CFRP with randomly aligned chopped carbon fibers are deteriorated more than expected, due to the decreased aspect ratio of the fibers after extrusion and injection molding[1)]. Long fiber (reinforced) thermoplastic (LFT) technology has been stressed as alternative to solve such the problem. It is expected that LFT technology may provide increased stiffness, strength and impact toughness, compared to conventionally extrusion/injection processed CFRP[2)]. An acrylonitrile-butadiene-styrene (ABS) resin is industrially important and versatile in automobile and other applications because of its high toughness, dimensional stability and good surface texture[3)]. It is thought that incorporation of carbon fiber into ABS resin as reinforcement by means of LFT technology may contribute to their individual benefits in terms of mechanical and thermal properties, toughness, etc. The objectives of the present study are to produce carbon fiber/ABS LFT pellets with different fiber lengths and to investigate the effect of the length of the LFT pellets on the mechanical properties of the composites.

## MATERIALS AND METHODS

### Material

In this work, carbon fiber/ABS LFT pellets were produced using continuous carbon fiber (PAN-based, 12k, Toray AMK, Korea) and commercial ABS pellets (Kumho Petrochemical Co., Korea) by using the LFT machine. Fig. 1 presents a schematic of LFT processing machine combined with extrusion and pultrusion concepts.

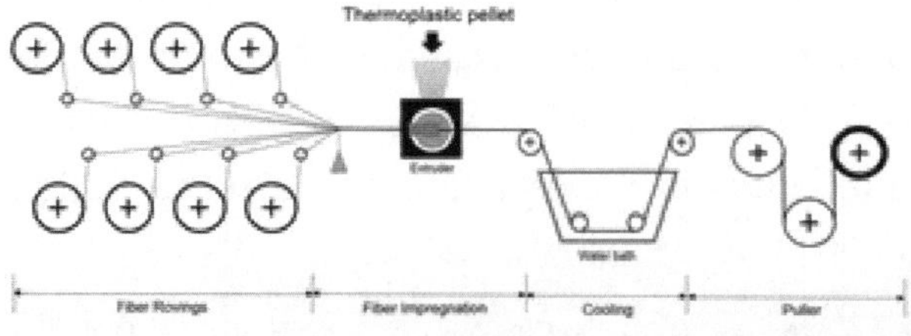

**Fig. 1.** Schematic of a processing machine to produce the LFT pellets.

The LFT pellets with different fiber lengths were directly used for injection molding to produce the carbon fiber/ABS composites. The average lengths of the obtained LFT pellets were 6, 12, and 24 mm. Tensile, flexural, impact, and dynamic mechanical properties and heat deflection temperature of the composites are investigated.

## RESULTS AND DISCUSSION

Fig. 2 compares the three-point flexural strengths of neat ABS, CF/ABS composites with (S-6, S-12) and without extrusion process (L-6, L-12, L-24). Here, S and L designate short (chopped) carbon fibers and LFT pellets, respectively. The flexural strength of all the composites is considerably higher than that of neat ABS. The strength of the composites made with LFT pellets is higher than that made with short carbon fibers. The composite (L-12) exhibited the highest flexural strength among the specimens measured. The possible reason for that is that the length of carbon fibers being distributed in the ABS matrix of the CFRP retains longer with the LFT pellets obtained without extrusion process.

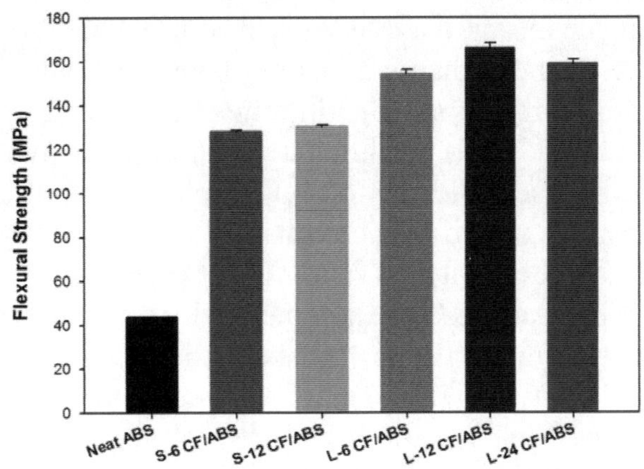


**Fig. 2.** A comparison of the flexural strengths of carbon fiber/ABS composites produced with different LFT pellet lengths.

## CONCLUSIONS

It has been found that the properties of the thermoplastic CFRP produced through LFT technology in the absence of extrusion process are higher than those of the CFRP produced including extrusion process, depending on the LFT pellet length. However, due to the limitation of the current LFT facility the impregnation level of melt-extruded ABS resin into the continuous carbon fiber tow is not efficient enough. So, further processing studies should be performed to have uniform and efficient impregnation attainable through the LFT process and also further increase the mechanical and thermal properties of the resulting composites.

## REFERENCES

1. Novo, P. J., Silva, J. F., Nunes, J. P., Marques, A. T. (2016). "Pultrusion of fibre reinforced thermoplastic pre-impregnated materials", *Composites Part B*, Vol. 89, p.328-339.

2. Luo, H., Xiong, G., Ma, C., Li, D., Wan, Y. (2014). "Preparation and performance of long carbon fiber reinforced polyamide 6 composites injection-molded from core/shell structured pellets", *Mater. Design*, Vol. 64, p. 294-300.

# Production of Cone-Shaped Graphitic Whisker in Scrolled Natural Polymer Film as "Microreactor"

Yukie Saito[1*], Valeriy Luchnicov[2]

[1]Graduate School of Agricultural and Life Sciences, The University of Tokyo, Tokyo, 113-8657, Japan.
[2]Institute of Material of Science of Mulhouse (IS2M), French National Center for Scientific Research (CNRS) , Mulhouse, 68057, France.
*aysaito@mail.ecc.u-tokyo.ac.jp

## INTRODUCTION

Cone-shaped graphitic whiskers (CGW) are a type of carbon whiskelrs, characterized by a super-helical cone stack of graphene[1)]. CGW are easily produced by carbonization of woody plants[2)], particularly of silica-rich species, because silica can be converted to SiC via heat treatment at >2000°C, which works as nuclei for a stable growth of CGWs with their apex angles in the range of 135-140°. However, the growth mechanism has not been elucidated. Our previous study[2)] indicated that wood cells work as "microreactors" affecting the shape and size of the CGWs grown inside. The wood cell is considered to have multiple functions, namely, to pyrolytic gas, accumulate it inside, and simultaneously work as a scaffold for the SiC nuclei. Anatomically different species of woody and herbal plant cells imply that differences in the size, permeability of gas, and Si deposition affected the features of the resultant CGWs[2)]. To elucidate the growth mechanism, we need to prepare for man-made "microreactors" wherein the cavity size can be controlled.

A type of "intelligent" polymer films has been invented, which self-scroll responding to circumstances because of their expansion-gradient in the thickness direction[3)]. By designing the thickness and gradient of the film, the curvature of scrolling can be controlled. Recently, self-scroll systems have been successfully produced by bio-polymers[3,4)], which fairly maintain their original shape under the carbonization process.

In this study, we aimed to use this biopolymer scroll as a "microreactor" and synthesize CGWs inside, under controlling.

## MATERIALS AND METHODS

### Preparation of the chitosan tubes

A solution of chitosan in acetic acid was casted on a glass substrate and a solid film was prepared. The film casted on the glass was immersed in sodium tripolyphosphate (TPP) solution and gradient TPP cross-linking was performed in the direction of the thickness of the film. The dried film was cut in small pieces, detached from the glass substrate and deposited on the surface of water, and scrolled to a tube with a diameter of several decades of μm.

### Carbonization

The resultant chitosan tubes were connected to a charcoal block, so that the carbon source gas is supplied from the charcoal during heat treatment. They were placed in a carbon crucible and heated at 2450°C for 1 h under Ar atmosphere.

## RESULTS AND DISCUSSION

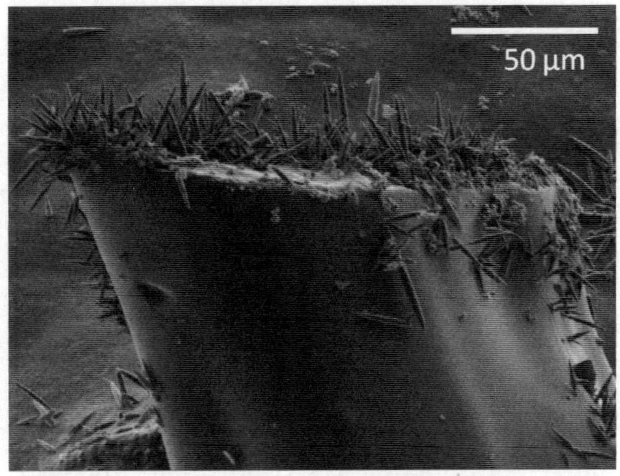


Fig. 1. SEM image of the CGWs grown in an end of the carbonized chitosan tube.

The microtubes of chitosan maintained their cavity even after heat treatment at 2450°C, although those made of polystyrene (PS) - poly(4-vinyl pyridine) (P4VP) lost the cavity during the heat treatment. In general, compared with synthetic polymers, natural polymers, such as chitosan and cellulose, show higher shape stability of the shape under carbonization Therefore, to function the tube as a microreactor for carbonization, natural polymer tubes may suit.
CGW s were observed on the surface of a tube, indicating the accessibility of nuclei Si for the carbonized TPP-crosslinked-chitosan. Notably, the growth of CGWs was especially concentrated on the end of the tube (**Fig. 1.**), probably a "drift" was generated toward the edge wherein more carbon was deposited.
The fractured surface of a carbonized chitosan microtube revealed that CGWs of a few µm diameter were successfully grown inside the cavity (**Fig. 2.**). By designing and controlling the shape, size and length of the cavity, further studies can be undertaken to elucidate the function of micro-spaces and nuclei for the growth mechanism, which affects the yield and structural varieties of the CGWs.

## CONCLUSIONS

A microtube was prepared by self-scrolling of a chitosan film with gradient TPP cross-linking. CGWs with diameters of a few µm were grown in the cavity of the chitosan microtubes by heat treatment at 2450°C with Si as nuclei and pyrolyzed gas from wood char as the carbon source. By further controlling the cavity shape of the microtube, the function of the "micro-space" and the growth mechanism of CGWs will be elucidated.

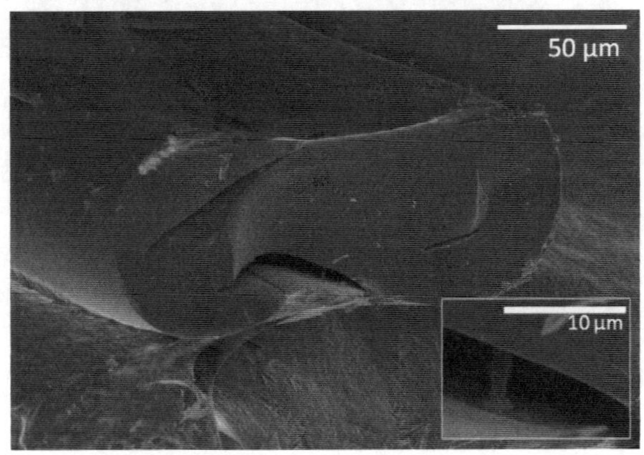


**Fig. 2.** SEM image of a fractured surface of the carbonized chitosan tube and a CGW grown inside the cavity (inset).

## REFERENCES


1. Saito, Y., and Nishio-Hamane, D. (2016). "Helical superstructure of continuum graphene cone uncovered by TEM analysis of herringbone-striped pattern in graphitic whiskers", *J. Crystal Growth*, Vol.451, p.27-32
2. Saito, Y., and Arima, T. (2007). "Features of vapor-grown cone-shaped graphitic whiskers deposited in the cavities of wood cells", Carbon, Vol.45, p.248-255.
3. Luchnikov, V. A., Saito, Y. and Tzanis, A. (2012). "A fibrous material created by self-rolling of a patterned polymer thin film", *Macromol. Rapid Commun.*, Vol.33, P.1404-1408
4. Egunov, A. I., Inaba, A., Gree, S., Malval, J.-P., Tamura, K., Saito, Y. and Luchnikov, A. V. (2016). "Time-programmed release of fluoroscein isocyanate dextran from micro-pattern-designed polymer scrolls", *J. Control. Release*, Vol.233, p.39-47

## Progressive Fabrication of Graphene and their Potential Applications

Won-Chun Oh*

*Department of Advanced Materials Science & Engineering, Hanseo University, Chungnam 356-706, South Korea*

(Email: wc_oh@hanseo.ac.kr)

## INTRODUCTION

Graphene has extraordinary physical and electrical properties, so that it has attracted great interest in different scientific areas in the last years. While graphene is a nanomaterial with high electron mobility and elevated conductivity, it has been reported that bare graphene, graphene oxide with new synthesis methods or the combination with other semiconductor materials can be used as catalyst materials. We examined the photocatalytic activity and catalytic recyclability with some kinds of semiconductor/graphene (binary type) or semiconductor/graphene-$TiO_2$ (ternary type) nanocomposites fabricated with modified chemical and physical methods. The prepared composites were characterized by some kinds of spectroscopic techniques including transmission electron microscopy (TEM) and Raman spectroscopic analysis These graphene-based photocatalysts has a great promise for various material/device applications, including solar cells, standard dyes and commercial dyes decomposition and $H_2$ evolution.

## METHODOLOGY

Firstly, graphene-$TiO_2$ was synthesized according to a facile solvothermal synthesis method [1]. Briefly, 40 mg of the graphene oxide (GO) synthesized by modified Hummers' method was dispersed in 140 mL of glacial acetic acid in a flask. The flask was sealed and sonicated for 3 h, and then tetrabutyl titanate was dropwise added into the above GO suspension under magnetic stirring. The mixture was stirred and then transferred into a Teflon lined stainless steel autoclave heated at 180 ◦C for 24 h. The obtained precipitate was washed 5 times with ethanol, and dried in vacuum overnight.

## RESULTS AND DISCUSSION

According to the XRD pattern of graphene, the predominant peak at $2\theta = 25.11$ corresponds to the (002) plane of graphene, which is characteristic of graphene. For TiO2, the characteristic peaks at $2\theta = 25.34$, 37.78, 48.14, 53.98, 62.68, and 68.98, which correspond to the anatase planes of (101), (004), (200), (105), (211), and (204) respectively, indicating that the synthesized $TiO_2$

catalyst existed in an anatase phase (JCPDS No. 21-1272) [2]. The morphology of the as-prepared photocatalysts of graphene-$TiO_2$ was investigated by the TEM. The size of the graphene-$TiO_2$ particles ranged from 250 to 350 nm, and the particles tended to agglomerate. We carried out Raman spectroscopy experiments for further investigation of structural phases of anatase graphene-$TiO_2$ nanocomposites. Raman spectra of anatase graphene-$TiO_2$ nanocomposites exhibit two characteristic Raman bands, D band at 1350 $cm^{-1}$ and G band at 1600 $cm^{-1}$. The D band is assigned to edge or in-plane $sp^3$ defects and disordered carbon, whereas the G band arises from the inplane vibration of ordered $sp^2$-bonded carbon atoms.

## CONCLUSION

Firstly, graphene-$TiO_2$ photocatalyst was synthesized according to a facile solvothermal synthesis method. Graphene-$TiO_2$ showed higher surface area and narrow band gap energy. In addition, the graphene-$TiO_2$ nanocomposites facilitate the separation of photogenerated charge carriers along with high stability, which ultimately enhance the photocatalytic activity under ultraviolet irradiation. So, these graphene-based photocatalysts has a great promise for various material/device applications, including solar cells, standard dyes and commercial dyes decomposition and $H_2$ evolution.

**References:**

1. H. M. Yadav, J.S. Kim, *Journal of Alloys and Compounds*, Volume 688, July 2016, Page 123–129.

[3] R. Sahraei, Z. S. Pour, M. Ghaemy, *Journal of Cleaner Production,* Volume 142, January 2017, Page 2973-2984.

## Radial Distribution Analysis of Carbon Fibers using Neutron Diffraction

Kiminori Ono[1*], Haruki Okuda[1], Masaru Nakada[2], Fumihiko Tanaka[1]

[1] Composite Materials Research Laboratories, TORAY Industries Inc., Ehime, Japan
[2] Toray Research Center, Inc., Shiga, Japan
*Kiminori_Ono@nts.toray.co.jp

### INTRODUCTION

Polyacrylonitrile(PAN)-based carbon fibers (CFs) have high Young's modulus and tensile strength owing to be consisted of both crystalline and amorphous phases [1]. The crystallite size and the orientation of the CFs affect those mechanical properties. The molecular dynamics simulation suggests that the basal plane of CFs has some vacancy defects [2]. Although those atomic arrangements would affect the mechanical properties, no experimental evidence to support those atomic arrangements of CFs was provided. The crystallite size has been analyzed by X-ray diffraction (XRD) analysis or high resolution transmission electron microscopy (HRTEM). The crystallite size analyzed by XRD is the planar part of the aromatic portion, while the size analyzed by HRTEM includes non-planer nature [3]. On the other hand, neutron diffraction technique provides the diffraction of atoms, while X-rays interact with the electron cloud, providing precise information on the radial distribution. In the present study, the radial distribution analysis of CFs, which have anisotropic structures, has been studied using neutron diffraction technique to investigate atomic arrangements, which would be important role of Young's modulus and tensile strength of CFs.

### MATERIALS AND METHOD

In order to examine the different types of basal plane, we prepared high modulus carbon fibers (HMCF), intermediate modulus carbon fiber (IMCF) and pitch carbon fibers (Pitch CF). The diffraction data was measured using iMATERIA at J-PARC (BL20). Neutron beam was irradiated to fiber axis from 45° to obtain the diffraction of meridional and equational direction because CFs have anisotropic structures. The scattering vector $Q$ was obtained up to 60 Å. The radial distribution function (RDF), including the atomic-scale information, was obtained by the following

$$RDF(r) = 4\pi\rho r^2 \{1 + \frac{1}{2\pi^2 \rho r} \int Q[S(Q) - 1]\sin(Qr)dQ\} \quad (1)$$

where $S(Q)$ is structure factor, $\rho$ is average number density of each CF. XRD experiments were performed with a laboratory X-ray diffractometer (PANalytical, X'Pert PRO) operated at 45 kV and 40 mA with CuK$_\alpha$. The crystallite size in-plane ($L_a$) and in stacking direction ($L_c$) were estimated by applying the Scherrer equation. Scherrer constant depending on the shape and dimensionality of the crystallites and is for turbostratic carbon materials given by $K = 0.91$ for $L_c$ and $K = 1.84$ for $L_a$.

### RESULTS AND DISCUSSION

The RDF of meridional direction of HMCF and Pitch CF indicated the peak toward long radial distance. On the other hand, the RDF of meridional direction of IMCF was attenuated around 18-21 Å, which was lower than 48 Å of the crystallite size ($L_a$) measured by XRD. This suggests IMCF has large distribution of the crystallite size. **Figure 1.** shows the RDF of meridional direction of HMCF, IMCF and Pitch CF around 1-5 Å. The radial distances after 2.43 Å of the next nearest neighbor of IMCF were 20% lower than those of HMCF. Carbon atoms around 2.43 Å of IMCF would be 20 % lower than HMCF, suggesting that IMCF has lattice defect compared to HMCF.

The RDF of equatorial direction of HMCF and Pitch CF indicated the periodic peak of 3.4-3.7 Å, suggesting that the stacking of the crystallite. On the other hand, the RDF of equatorial direction of IMCF was attenuated around 15 Å, which was lower than 19 Å of the crystallite size ($L_c$) measured by XRD. This suggests IMCF has large distribution of the $L_c$. **Figure 2.** shows the RDF of meridional direction of HMCF, IMCF and Pitch CF around 1-3 Å. The peaks of equatorial direction of IMCF had shoulder of 2.28, 2.54, and 3.17 Å compared to HMCF. These shoulder peaks of distances are equivalent to the radial distance of 5-memberd ring and 7-memberd ring, suggesting that IMCF would include 5-memberd ring and 7-memberd ring. These results of RDFs suggest that atomic arrangements are important role of Young's modulus and tensile strength of CFs.

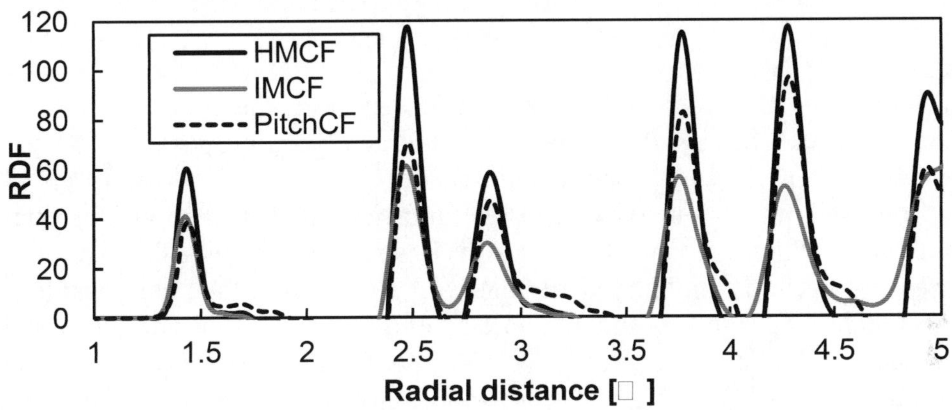


**Fig. 1.** The RDF of meridional direction of HMCF, IMCF and Pitch CF

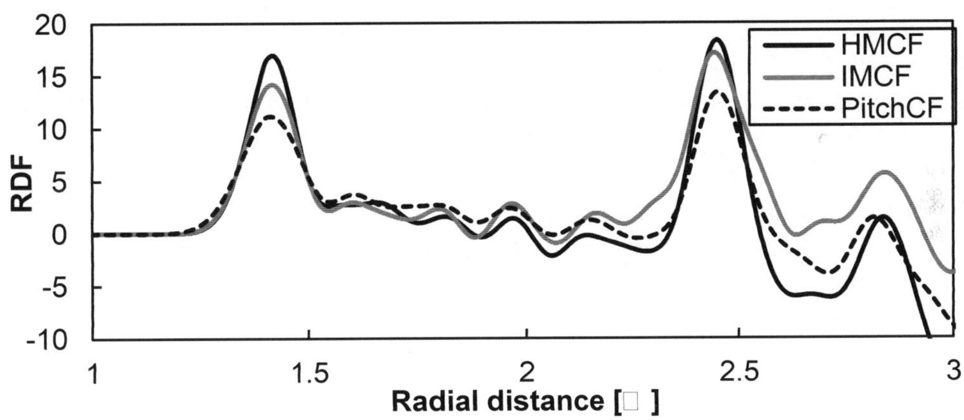


**Fig. 2.** The RDF of equatorial direction of HMCF, IMCF and Pitch CF

## REFERENCES


[1] F. Tanaka, T. Okabe, H. Okuda, I.A. Kinloch, R.J. Young, The effect of nanostructure upon the compressive strength of carbon fibres, Journal of Materials Science 48(5) (2013) 2104-2110.

[2] A. Ito, S. Okamoto, Atomistic structure of a PAN-based carbon fibre constructed by molecular dynamics simulations and lifetime measurement of positron annihilation, International Journal of Sustainable Aviation 1(3) (2015) 269-287.

[3] H. Aso, K. Matsuoka, A. Sharma, A. Tomita, Structural analysis of PVC and PFA carbons prepared at 500–1000 °C based on elemental composition, XRD, and HRTEM, Carbon 42(14) (2004) 2963-2973.

# Regioselective Catalysts for Repair of Graphene Structure from Graphene Oxide

Jurgis Barkauskas[1]*, Justina Gaidukevič[1], Ieva Mališauskaitė[1], Rasa Pauliukaitė[1,2], Gediminas Niaura[2], Ieva Matulaitienė[2], Gvidas Astromskas[2], Virginijus Bukauskas[2], Romualdas Trusovas[2]

[1]Vilnius University, Faculty of Chemistry and Geosciences, Vilnius, Lithuania.
[2]Center for Physical Sciences and Technology, Vilnius, Lithuania
*jurgis.barkauskas@chf.vu.lt

## INTRODUCTION

The structure of GO, different from graphene, has been the subject of discussion for many years. Several models considering the structure of GO have been proposed[1)]. The stoichiometry of oxygen functionality in GO batches vary depending on the preparation conditions. Therefore, despite the recent progress in the chemistry of GO, the question either GO should be considered as a family of materials or a single composition is still under discussion. Likewise, the mechanisms of the processes that occur during the reduction of GO are far from the precise understanding[2)].

The basal plane of graphene (and, to some extent, that of GO) has an increased electron density[3)]. In consequence, it should be attacked by an electrophile, the particle with a lack of electron density. On the contrary, nucleophiles will tend to attack the atoms above and below the basal plane of graphene. The use of appropriate reagents combined with a proper preparation protocol (e.g., laser treatment) enable to develop a tuneable control of the reduction of GO. A considerable amount of literature has been published on the reduction of GO in an aqueous medium, taking into account that the water molecule is a nucleophile. Meanwhile, the same process in an electrophilic media has been studied in much less extend. The aim of this research was to investigate the process of GO reduction in the presence of $H_3BO_3$, since the boron atom is regarded as essentially electrophilic in character. Moreover, in the presence of polyols (e.g., glycerol), $H_3BO_3$ is able to form a chelate complex, which can act as a catalyst in epoxide ring opening reaction[4)]. The epoxide ring opening reactions in the presence of $H_3BO_3$ are highly regioselective. The epoxide groups are highly abundant on the basal planes of GO[1)].

## MATERIALS AND METHODS

The process of GO reduction in the melt of pure $H_3BO_3$, $H_3BO_3$ + glycerol and $H_3BO_3$ + ascorbic acid has been studied up to 400 °C by means of thermogravimetry/differential scanning calorimetry (TG/DSC). Furthermore, the mechanisms of the reactions was examined by using a laser excitation method. The synthesized products (electro-active materials, EAMs) were analyzed using a scanning electron microscopy (SEM), X-ray diffraction (XRD), Raman spectroscopy and Fourier transformed infrared spectroscopy (FTIR). Furthermore, density and electrical conductivity measurement was performed. Specific surface areas and pore volume distribution was evaluated from nitrogen adsorption isotherms. The electrochemical characteristics were investigated using the cyclic voltammetry (CV) and electrochemical impedance spectroscopy (EIS) methods.

## RESULTS AND DISCUSSION

The main conclusion to be drawn from TGA/DSC study is that the melt of $H_3BO_3$ is suitable for the exfoliation and modification of GO. The additives that form complexes with the boron atom (ascorbic acid and glycerol) influence the process of exfoliation and can be used to achieve a more tunable structure of the EAM. The kinetics of the process of GO reduction in the melt of $H_3BO_3$ is rather different from that in aqueous media and thermal reduction. The TGA/DSC analysis has also revealed that to achieve a tunable structure of EAM, the melting with $H_3BO_3$ is superior in comparison with the thermal treatment. Furthermore, the process has been studied using a laser excitation method. It was optimized by choosing the excitation wavelength and duration.

The results of XRD analysis indicate that the crystalline structure of graphite in EAMs is restored to some extent after the reduction of GO in the presence of $H_3BO_3$. No significant differences are observed in the XRD patterns after the thermal treatment of EAMs up to 800 °C; it is evidence of their high thermal stability.

Raman and FTIR spectroscopy show that the amount and variety of defects and functional groups is substantially diminished through the procedure of reduction of GO. It is worth noting that the variety of defects and functional groups that remain after the reduction substantially depends on the presence of $H_3BO_3$. This confirms the earlier hypothesis about the regioselective catalysis of boron-containing complexes.

The results of density and electrical conductivity measurement revealed that the granulometric composition of EAMs is comparable, while electrical conductivity varies about tenfold. The measurement of pore size distribution indicated the presence of well-developed mesopores uniform in diameter (4 nm) for all EAMs. The conclusion from the studies carried out suggests that the synthesized EAMs should be suitable for supercapacitors.

CV and EIS were used to determine double layer capacitance of the EAM samples. The results showed that reduction conditions are extremely important for the capacitance values. EIS gave more precise evaluation of the specific double layer capacitance and it was in the range up to 1 mF $cm^{-2}$. Most probably it is due to changes of the porosity of the film.

## CONCLUSIONS

The process of GO reduction in the presence of $H_3BO_3$ has been investigated. $H_3BO_3$, different from $H_2O$, is an electrophilic reagent, which should attack the basal plane of GO and act as a regioselective catalyst in epoxide ring opening reaction. The study of kinetics and the structure of synthesized products confirm that the boron-containing complexes enable to control the number and variety of structural defects in reduced GO. The double layer capacitance og synthesized samples was determined by means of CV and EIS measurements.

## ACKNOWLEDGEMENT

This research is funded by Lithuanian Research Council via measure "Towards Future Technologies", Project No. LAT 11/2016.

## REFERENCES

1. Dreyer, D. R., Todd, A. D., Bielawski. C. W. (2014). "Harnessing the chemistry of graphene oxide", *Chem. Soc. Rev.*, Vol.43, p.5288-5301.

2. Pei, S. Cheng, H.-M. (2012). "The reduction of graphene oxide", *Carbon*, Vol.50, p. 3210-3228.

3. Russo, P., Hu, A., Compagnini, G. (2013). "Synthesis, properties and potential applications of porous graphene: a review", *Nano-Micro Lett.*, Vol.5(4), p.260-273.

4. Halimehjani, A. Z., Gholami, H., Saidi, M. R. (2012). "Boric acid/glycerol as an efficient catalyst for regioselective epoxide ring opening by aromatic amines in water", *Green Chem. Lett. Rev.*, Vol.5(1), p.1-5.

# Role of Porosity in the Hygroscopic Nature of Nanodiamonds

Elda-Zoraida Piña-Salazar[1,2], Koki Urita[3], Takuya Hayashi[4], Eiji Osawa[5], Toshio Sakai[6], Katsumi Kaneko[2,*]

[1] Interdisciplinary Graduate School of Science and Technology, Shinshu University, Nagano, Japan
[2] Center for Energy and Environmental Science, Shinshu University, Nagano, Japan
[3] Department of Applied Chemistry, Faculty of Engineering, Nagasaki University, Nagasaki, Japan
[4] Department of Water Environment and Civil Engineering, Shinshu University, Nagano, Japan
[5] Nano-Carbon Research Institute, Ltd., Ueda, Nagano, Japan
[6] Department of Materials Chemistry, Faculty of Engineering, Shinshu University, Nagano, Japan
* kkaneko@shinshu-u.ac.jp

## INTRODUCTION

Detonation nanodiamonds have been widely used. They are aggregate structures (50-500 nm) of polyhedral particles of $sp^3$ hybridized carbon atoms as small as 4-5 nm in diameter,[1] such material possess an inert core and a chemically active surface. Aggregates of nanodiamonds have inter-granular gaps, which account mainly for mesoporosity.[2] Nanodiamonds have the ability to contain water in their pores formed upon aggregation.[3] Wettability to water indicates its affinity to water. It can be described by vapour water adsorption and hygroscopic tests. So far today, there have been few attempts to analyze deeply the adsorption of water and wettability of nanodiamonds especially with the relevance to their wide range of structures. Then, this work analyzed the dependency of such properties on their hierarchical porosity.

## MATERIALS AND METHODS

Porosity of nanodiamonds (hard hydrogel; Nano-Carbon Research Institute, Ltd., Japan) was determined by $N_2$ adsorption at 77 K after preheating at 423 K and $10^{-4}$ Pa for 3 h. Afterward, wettability was evaluated by water vapor adsorption isotherm at 298 K after pre-treating at 423 K and 2 mPa for 2 h, and by absorption of liquid water by dropping droplets of 6.5 μL of water on nanodiamonds pellets. Complementary analysis using Electron Microscopy (SEM), High-Resolution Transmission Electron Microscope (HR-TEM), Electron Energy Loss Spectroscopy (EELS), X-ray diffraction pattern (XRD) and Temperature Programed Desorption coupled with a Mass Spectrometer (TPD-MASS) elucidated the association among the characteristic chemistry and pore structures with the wettability of nanodiamonds.

## RESULTS AND DISCUSSION

The whole feature of water adsorption isotherm of highly crystalline detonation nanodiamonds (XRD) **(Fig. 1)**, is a characteristic Type V isotherm according to IUPAC classification. This indicates that nanodiamonds are weak hydrophobic porous material.[4] The slight adsorption of vapor water in the low relative pressure range, suggests the presence of hydrophilic sites on nanodiamonds where first water molecules are adsorbed.

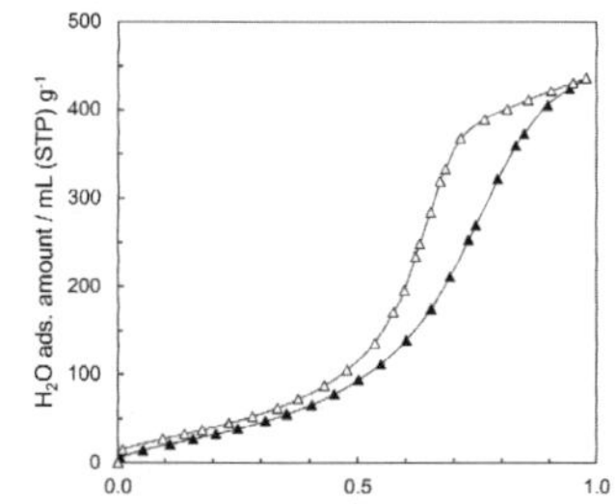


**Fig. 1.** Water adsorption isotherm of nanodiamonds at 298 K.

Functionalities are located in the vicinity of the pores situated on the graphitic shells that surround nanodiamonds cores, as proven by ELLS ($sp^2$ hybridized carbon on the outer layer) and TPD profiles (desorption of $CO$ and $CO_2$ derived from the oxygen containing-carbon groups or amorphous carbon situated on the graphitic surface that covers the nanodiamond core). Such functionalities in those pores and curvature of such diminutive cavities allow the attraction of water molecules. Subsequently, clusters of water grow inducing condensation in narrow pores to fill nanodiamond voids.[5]

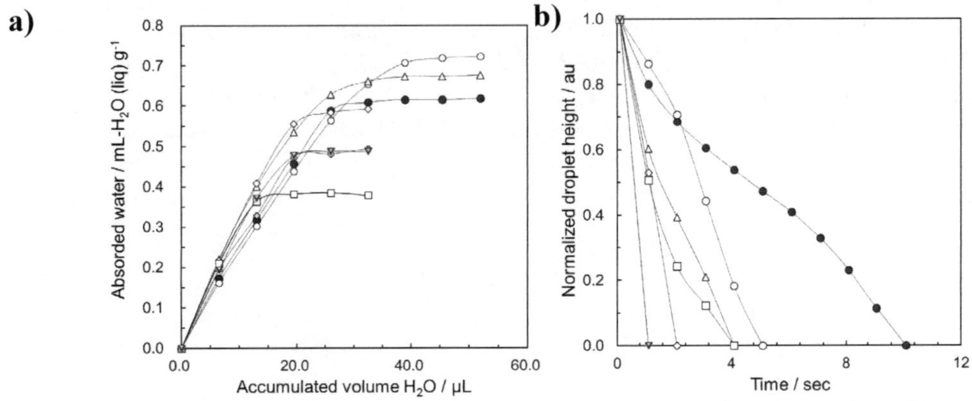


**Fig. 2.** Absorption of water droplets (a) and kinetics of water droplet absorption (b) on pelletized porous materials; nanodiamonds (●(1.6 g $cm^{-3}$)) and (○ (1.4 g $cm^{-3}$)), montmorillonite (Δ(1.3 g $cm^{-3}$)), mordenite (□(1.0 g $cm^{-3}$)), ZSM-5 (□(1.1 g $cm^{-3}$)), molecular sieve 5A (□(1.2 g $cm^{-3}$)) and silica gel (□(1.1 g $cm^{-3}$)).

Measurements of hygroscopicity revealed that nanodiamond pellets absorbed water droplets up to 0.73 mL (liq.) $H_2O$ $g^{-1}$ (filling~84% void space), which was comparable to 0.68 mL(liq.) $H_2O$ $g^{-1}$ (filling~67% void space) for mesoporous montmorillonite, and at least 20% higher than that of microporous mordenite, ZSM5, molecular sieve 5A and silica gel (**Fig. 2**). Interestingly, in spite of the hydrophobicity presented at the molecular level, nanodiamond possess intensive hygroscopic property; even when the speed of water droplets absorption of the other porous materials is higher. Here the presence of micropores triggers the adsorption of water molecules; and micropores and small mesopores (< 6 nm) promote higher absorption of water.

## CONCLUSIONS

The small porosity of nanodiamonds, as well as the hydrophilic sites presented in the nanodiamond shell, are key factors to confer them special wettability as was proven here because oxygen-containing functional groups localized in those pores and curvature of such diminutive cavities allow the attraction of clusters of water molecules that fill these voids.

## REFERENCES


1. Ozawa, M., Inaguma, M., Takahashi, M., Kataoka, F., Krüger, A. and Osawa, E. (2007). "Preparation and Behavior of Brownish, Clear Nanodiamond Colloids", *Adv. Mater.*, *19* (9), p. 1201–1206.
2. Kondo, T., Kobayashi, M., Saito, T., Kadota, Y., Kameshima, T., Aikawa, T., Kawai, T. and Yuasa, M. (2014). "Micrometer-Sized Mesoporous Diamond Spherical Particles", *Diam. Relat. Mater.*, *43*, p. 72–79.
3. Batsanov, S. S., Gavrilkin, S. M., Batsanov, A. S., Poyarkov, K. B., Kulakova, I. I., Johnson, D. W. and Mendis, B. G. (2012). "Giant Dielectric Permittivity of Detonation-Produced Nanodiamond Is Caused by Water", *J. Mater. Chem., 22* (22), p. 11166–11172.
4. Thommes, M., Kaneko, K., Neimark, A. V., Olivier, J. P., Rodriguez-Reinoso, F., Rouquerol, J. and Sing, K. S. W. (2015). "Physisorption of Gases, with Special Reference to the Evaluation of Surface Area and Pore Size Distribution (IUPAC Technical Report)", *Pure Appl. Chem., 87* (9–10), p. 1051–1069.
5. Kaneko, K. (2015). "Graphitic Nanopores: Water Capture in Carbon Cuboids", *Nat. Chem. 7* (3), p. 194–196.

# Self-assembly of Nitrogen-doped Carbon Nanobelts via Catalytic Chemical Vapour Deposition

Juan L. Fajardo-Diaz[1], Alejandro J. Cortés-López[1], Florentino López-Urías[1], Emilio Muñoz-Sandoval[1*]

[1]Advanced Materials Division, IPICYT, San Luis Potosí, México.
*ems@ipicyt.edu.mx

## INTRODUCTION

Efforts to synthesize a self-assembled structures have been made through the recent years using nano-building blocks in a "bottom up" synthesis [1,2]. In the case of complex self-assembled carbon nanostructures different methodologies are used like sol-gel solutions [3], organic matrix decomposed via pyrolysis[4] and catalytic chemical vapour deposition. Nevertheless few papers reports a self-assembly mechanism during the production of carbon nanostructures. In this work we use a catalytic aerosol assisted chemical vapour deposition system using cobalt-copper thin films as catalyst and benzylamine, thiophene and ferrocene as precursor to obtain carbon nanofibers that works as "building blocks" through the formation of carbon nanobelts. As result we found the formation of cobalt nanoparticles with size ~200 nm, which are covered by copper and copper oxide in a core-shell configuration $Co@Cu_xO$. Also the presence of "monoliths" of ~2 mm length and 200 nm thickness that are conformed of copper where cobalt nanoparticles agglomerates and could be consider as the catalyst particle of the carbon nanobelts. The carbon belts are built from wrinkled and corrugated carbon nanofibers with a high disorder in their graphitic sheets ($I_D/I_G > 1$) and with pyrrolitic nitrogen doped configuration that are criss-crossed.

## MATERIALS AND METHODS

The synthesis of the cobalt-copper thin film used as catalytic support was carried by a modified Intercovamex V3 sputtering system where Si/SiO2 substrates were place on a holder inside of a high vacuum deposition chamber and a plasma process took place. For the deposition of cobalt thin film a DC MDX-500 power supply was used at 30W during 17 min and 7s to create a 20 nm film, in the case of copper a "Pinnacle plus" pulsed power supply at 15 W over 66 seconds to create a 2nm film thickness.

The production of the self-assembled carbon nanobelts was carried by aerosol assisted catalytic chemical vapour deposition using a two furnace system with different temperatures of 750 °C (inlet) and 850 °C (outlet) and one nebulizer where a mix of benzylamine, ethanol and ferrocene was used as carbon, nitrogen, oxygen and sulphur source. The $Si/SiO_2$ substrates containing the Co/Cu thin films were inserted in a quartz tube of 1.1 m length and 2.54 cm internal diameter and placed with a separation distance of 2 cm between each other. The reaction was performed during 2 hours maintaining the conditions all the time. The samples were taken from the quartz tube and characterized by scanning electron microscopy (SEM) and backscattered electron (BSE-SEM), high resolution transmission electron microscopy (HR-TEM), Raman spectroscopy with a laser wavelength of 532 nm, X-ray diffraction from 2θ of 10 to 90 degrees with a step size of 0.04 during 5 seconds, and X-ray photoelectron spectroscopy (XPS) with high resolution analysis over C1s, N1s, O1s.

## RESULTS AND DISCUSSION

**Fig. 1a** depicts the type of carbon nanostructures synthesized showing a diversity in the size and structure where two type of structures can be found: small concentration of cross-linked carbon nanofibers and high concentration of carbon nanobelts (CNBs). These CNBs structures are formed during the synthesis as part of a self-assembly effect where the cobalt-copper nanoparticles agglomerates in monolith structure (**Fig. 1b**) that catalyzes individual carbon nanofibers that interweave in one direction and create the CNBs. **Fig. 1c** shows a side-view of the CNBs with a 350 nm of thickness where wrinkled carbon nanofibers grows leaving a structure full of holes and defects. **Fig. 1d** illustrates an intermediate where small nanoparticles of Co (~200 nm) coalesce and creates small carbon belts.

**Fig. 2a** shows the structural conformation of the carbon nanofibers that works as a "building block" for the formation of the CNBs, this carbon nanofiber is full of defects, with a corrugated structure and can be compared with "wrinkled paper sheet" what make it that the structure contains holes and sharp edges, this characteristics promotes a self-assembly effect between the carbon nanofibers to build CNBs. **Fig. 2b** illustrates a section of one small CNBs where a carbon nanofiber is partially attached to the main structure. **Fig. 2c** shows a cobalt nanoparticle ~80 nm that catalyzes a nanofiber, this cobalt nanoparticle is surrounded by another type of crystalline material that by EDS analysis over the edge confirmed that contains copper and oxygen, what is believed that a "core-shell" configuration take place in a $Co@Cu_xO$, and is the reason why the cobalt particle don't catalyzes carbon nanotubes as the copper oxides works as grain boundary avoiding the continues growing of graphitic layers and leaving holes during its nucleation process.

**Fig. 3a** illustrates a Raman spectra of the CNBs where four signals were found part of the D band (1350.9 cm-1), G band (1582.8 cm-1), and the overtones 2D and D+G (2706.55 $cm^{-1}$ and 2927.98 $cm^{-1}$) with a behaviour similar to the reported for damaged graphene sheets. Also by the $I_D/I_G$ relation of 1.17 can be confirmed that CNBs is full of structural defects. In the case of the G band, a downshift could be related with a high concentration of sp3 probably as consequence of the nitrogen doping and influenced by the copper oxide. **Fig. 3b** reveals the N1s spectra of XPS characterization where can be seen that the CNBs are doped with nitrogen (1.8%), specifically a high concentration of pyrrolic nitrogen, followed by pyridinic-nitrogen, quaternary and n-oxide.

## CONCLUSIONS

In this work we found the conditions where the carbon nanobelts grows in one-step CVD synthesis. Reveal the influence of the copper in the growing of the carbon nanofibers and how this structures works as a unity for the formation of CNBs. The Raman characterization reveal how this material can be compared with disordered graphene sheets as the relation between the D band and G band is above 1, furthermore the XPS spectra reveals the presence of nitrogen in the structure.

## REFERENCES

1. S. Mann, (2009). "Self-assembly and transformation of hybrid nano-objects and nanostructures under equilibrium and non-equilibrium conditions", Nature Materials, vol. 8, p. 781-792.
1. D. J. Lunn , O. E. Gould, G. R. Whittell, D. P. Armstrong, K. P. Mineart, M. A. Winnik, R. J. Spontak, P. G. Pringle y I. Manners, (2016). "Microfibres and macroscopic films from the coordination driven hierarchical self-assembly of cylindrical micelles". Nature Communications, Vol.7.
2. S. Luo, Y. Luo, H. Wu, M. Li, L. Yan , K. Jiang, L. Liu, Q. Li, S. Fan y J. Wang, (2017). "Self-assembly of 3D Carbon Nanotubes Sponges: A Simple and Controllable Way to Build Macroscopic and Ultralight Porous Architectures". Advanced Materials, vol. 29, p. 8.

3. R. R. Kohlmeyer, M. Lor, J. Deng, H. Liu y J. Chen, (2011). "Preparation of stable carbon nanotube aerogels with high electrical conductivity and porosity". Carbon, vol. 49, p. 2352-2361.

## FIGURES

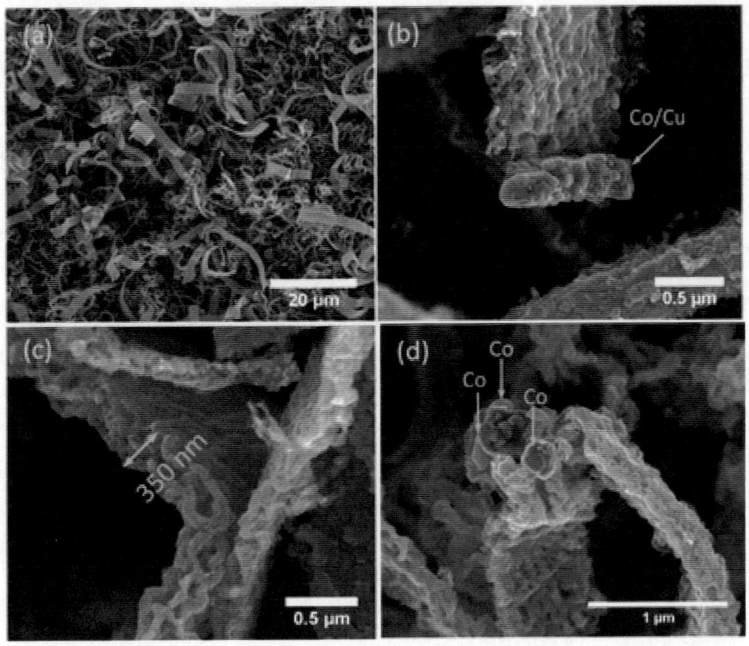


**Fig. 1:** Scanning electron micrograph of the CNBs: a) General view where can be observed a higher concentration of CNBs on the area. b) A close up over a agglomerated monolith of Co/Cu particles at the tip of the CNBs. c) Side edge of the CNBs with defects part of carbon nanofibers configuration. d) Agglomeration of cobalt particles catalyzing a small interweaving carbon nanofibers.

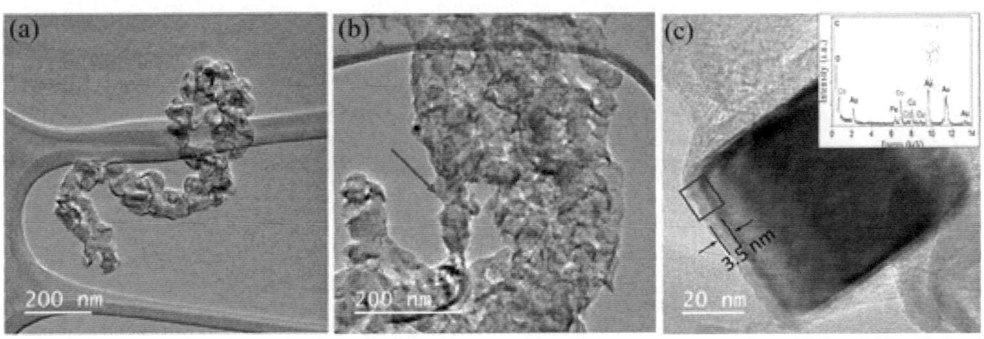


**Fig. 2**: Transmission electron micrograph of a) A unity like a "constructor block" of the CNBs, b) A section of the CNBs where a section of a carbon nanofiber is detached from the main structure. c) A cobalt particle that is surrounded by copper and copper oxide.

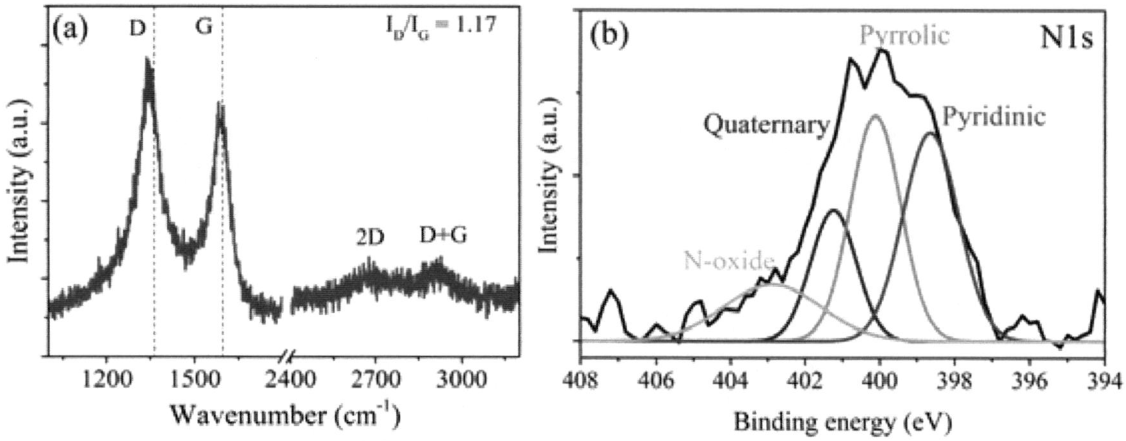


**Fig. 3:** a) Raman spectra of the CNBs structure where the relation of peak D and G is above 1. b) N1s Section of XPS spectra.

# Separation of Cone-Shaped Graphitic Whiskers by Density Gradient Centrifugation

Yuko Goto[1], Yukie Saito[1*]

[1]Graduate School of Agriculture and Life Sciences, The University of Tokyo, Tokyo, 113-8657, Japan.
*aysaito@mail.ecc.u-tokyo.ac.jp

## INTRODUCTION

Cone-shaped graphitic whiskers (CGWs) comprise highly ordered continuous hexagonal layers stacked in a helical–conical manner[1)]. CWGs are a type of pyrolytic carbon derived from carbonaceous materials heated >2000°C in the presence of nuclei, e.g., SiC, that can act as a trigger for stable growth with apex angles in the range of 135–140°, and they appear as whiskers of a few μm in diameter and a few decades in length[2)]. Although their unique physical properties can be attributed to their unique structure, some features such as adsorption and electro-chemical properties are not yet evaluated. One of the reasons is that pure CGWs cannot be produced in substantial amounts, and they are inevitably produced together with impurities, such amorphous carbon for the C gas source and SiC residues as nuclei. In this study, the separation of CGWs from impurities by density gradient centrifugation was studied. The idea was that CGWs can be trapped between the pair of solvent layers, the densities of which are either larger or smaller than that of the CGWs. The density of CGWs was previously calculated from the structural data from electron diffraction analysis and then exemplified by their sink-float behaviour in a solvent of adequate density.

## MATERIALS AND METHODS

### Material

Wood (*Quercus glauca*) char was powdered and treated at 2400°C under Ar atmosphere to produce the CGW samples, which include a certain amount of impurities. The density of a single CGW was calculated based on the data of inter- and intra-layer distances of the hexagonal carbon layers, derived from electron diffraction data.

### Density gradient centrifugation and estimation of the CGW ratio

In order to calculate the density, a series of solvents with densities neighbouring that of CGW was prepared by mixing organic solvents. The CGW sample, including impurities was added to the layered solvents comprising of higher and lower density and further centrifuged. The sample trapped between the pair of the layers was observed by scanning electron microscopy (SEM). The two-dimensionally projected areas of CGWs and impurities were measured as $A_c$ and $A_i$, respectively, and the CGW ratio, $A_c/(A_c+A_i) \times 100$ was calculated.

## RESULTS AND DISCUSSION

The density calculated from the electron diffraction data was 2.17 g/cm$^3$, although the apparent density measured by immersing in solvents of neighboring density was measured to be 1.9–2.0 g/cm$^3$, which is slightly lower than the value obtained by the crystallographic structure. Precise investigations using solvent-exchange methods revealed that the frill-like structure covering the surface of CGWs brought a buoyant force to a CGW and decreased the apparent density in the solvent.

For density gradient centrifugation, pairs of two layered solvents were employed (**Fig. 1.**). With any of them, in the middle band appearing between the layers of the lower and the higher density, the CGWs were selectively trapped. The highest CGW ratio was measured with the pair of 1.9–2.0 $g/cm^3$, and the value was as much as 40.3% (**Fig. 2.**). The ratio is apparently high compared to the original ratio of 6.0%, indicating the effectivity of the method.

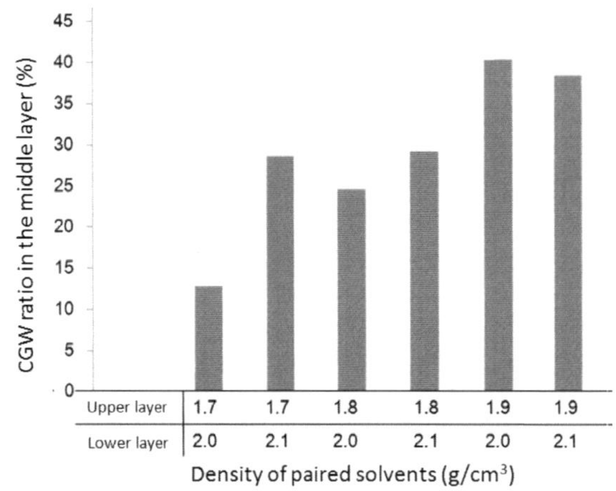


**Fig. 1.** CGW ratios observed in the middle layer appearing after centrifugation between the pair of solvents.

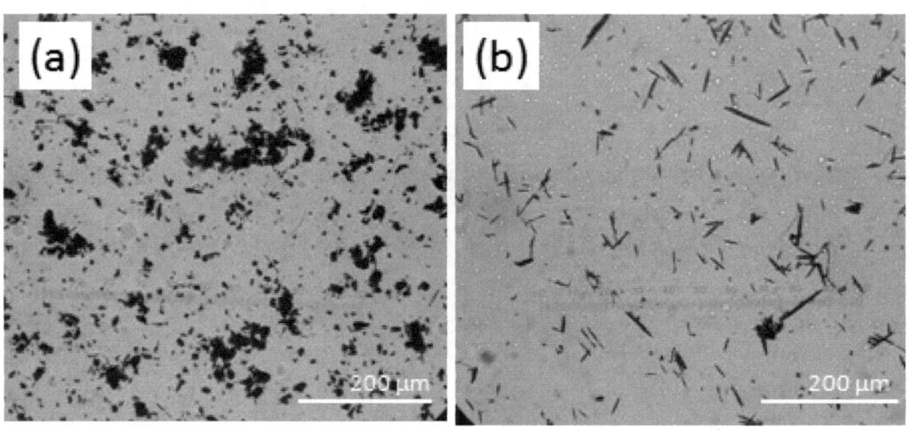


**Fig. 2.** Optical microscopy images of (a) original CGW samples with impurities and (b) CGWs selectively trapped in the middle layer

## CONCLUSIONS

The density gradient centrifugation method was used to successfully separate CGWs from impurities. This nondestructive method to enrich the CGW can be further applied for sample preparations for measurements of their unrevealed unique properties.

## REFERENCES

1. Saito, Y., and Nishio-Hamane, D. (2016). "Helical superstructure of continuum graphene cone uncovered by TEM analysis of herringbone-striped pattern in graphitic whiskers", *J. Crystal Growth*, Vol.451, p.27-32
2. Saito, Y., and Arima, T. (2007). "Features of vapor-grown cone-shaped graphitic whiskers deposited in the cavities of wood cells", Carbon, Vol.45, p.248-255.

# Effects of CNT Composite on Snow Sliding Property and Water Sliding Property on the Vibrated Surface of Silicone/ CNT Composite Sheets

Kenji Yanagisawa[1*], Mitsuki Tsuchiya[2]

[1] Department of Mechanical Engineering, Nagaoka University of Technology, Nagaoka, Japan
[2] Department of Mechanical Engineering, National Institute of Technology, Nagano College, Nagano, Japan.
*email@email.com

## INTRODUCTION

The lotus leaf has unique property of holding water on its surface without penetrating inside the leaf. This phenomenon is technically termed Lotus Effect or super-hydrophobicity. Super-hydrophobic surfaces are used not only for resisting water and condensation, but also for preventing contamination. Various devices with super-hydrophobic surface are developed. Super-hydrophobic surface can be produced using various methods such as sol-gel processing and chemical vapor deposition method[1,2)]. By these methods, it is difficult to produce hydrophobic surface in large-scale at low cost. Vapor grown carbon fiber (VGCF) is a new class of carbon fiber, which is distinct from other types of carbon fibers in its method of production[3,4)]. VGCF is of great scientific interest because of its excellent mechanical properties and electrical conductivity[5,6)]. Micro imprint performance of VGCF/polymer composites are examined and good transcription performance was achieved[7)]. And, studies on super-hydrophobic sheets by micro imprint methods were reported[8,9)]. Then, low drag on VGCF/ polymer composites surfaces was reported[10)]. Vibration has been used to accelerate the velocity of water sliding on a silicone sheet. So far, studies on velocity of snow sliding on silicone sheet were limited. If vibration has an effect on snow sliding on a silicone sheet surface, prevention of accidents caused by falling snow was expected. The aim of the present study was to investigate snow sliding property on the vibrated silicone sheet and the vibrated CNT composite sheet.

## MATERIALS AND METHODS

In this study, the surface functions are controlled with the shape and the alignment of the micro pattern [1], which can be machined in the following processes: 1) A thermosetting silicone resin is selected as matrix material for resin sheets. 2) Vapor Grown Carbon Fiber (VGCF) a good mechanical reinforcement potential to the plastic sheets. Snow particles were prepared in the following processes: 1) The snow particle size was unified by using sieve. Snow was crushed by ice crusher. 2) Silicone sheet was set on a tilted stage. 3) A snow particle was put on a silicone sheet. 4) A stage was vibrated by the vibrator. The vibrator was controlled by Amplifire. An acceleration sensor set to measure amplitude of vibrator. 5) Snow particle moving was taken with a video camera. 6) The moving distance of snow particle was measured after 100s.

## RESULTS AND DISCUSSION

The moving distance of water droplet and snow particles on the surfaces of the Silicone/VGCF sheets on tilting table are shown by Fig. 1. The white triangle shows the moving distance of water droplet on Silicone sheet and the black triangle shows the moving distance of water droplet on

Silicone/ VGCF sheets. The white bars and black bars show the moving distance of snow particles on Silicone and Silicone /VGCF sheets. The moving distance was increased for VGCF composite. The moving distance of snow particle was become longer than the moving distance of water droplet.

## CONCLUSIONS

In this study, we show the effects of VGCF composite on snow sliding property and water sliding property on the vibrated surface of Silicone/ CNT composite sheets. The velocity of water droplets and snow particles on Silicone/VGCF sheets was faster than the one on Silicone sheets for VGCF composite.

## REFERENCES


1. Shibuichi, S., Onda, T., Satoh, N. and Tsujii, K. "Super Water-repellent Surfaces Resulting from Fractal Structure," J. Phys. Chem. 100, 50, 1996, 19512-19517.
2. Li, Y., Liu, F. and Sun, J. "A facile layer-by-layer deposition process for the fabrication of highly transparent superhydrophobic coatings," Chem. Commun., 2009, 2730-2732.
3. Koyama, T., "Formation of carbon fibers from benzene," Carbon 10, 1972, 757– 758.
4. Koyama, T. and Endo, M., "Structure and growth processes of vapor-grown carbon fibers," Ohyo Butsuri 42, 1973, 690.
5. Tibbetts, G. G., Endo, M. and Beetz Jr., C. P., "Carbon fibres grown from the vapor phase: a novel material," SAMPLE Journal, 1986 September/- October, 30– 35.
6. Tibbetts, G. G., "Growing carbon fibers with a linearly increasing temperature sweep: experiments and modeling," Carbon 30, 1992, 399– 406.
7. Meguro, T., Song, S. M., Sugimoto, K., Hanaoka, M., Kawabe, S. and Osaka, H. "Micro/ Nano imprinting of PP/VGCF Nanocomposite Films with Highly Orientated Vapor Grown Carbon Nanofiber," Journal of the Japan Society for Precision Engineering 74, 2008 , 9, 960-964 (in Japanese).
8. Yanagisawa, K., Oshida, K., Song, S. M. and Sugimoto, K., Surface Pollution Control Technology of Photovoltaic System, CESEP'11, 2011, 118.
9. Yanagisawa, K., Song, S. M. and Sugimoto, K., Surface Pollution Control Technology of Resin/VGCF Composite Sheet with Textured Surface, International Tribology Conference Hiroshima 2011, 2011, PS2-01.
10. Yanagisawa, K., Song, S. M. and Sugimoto, K., Super-hydrophobic surface of Resin/VGCF Composite Sheet using Imprint Process, Carbon 2012, 2012, ID422.

# Solvothermal synthesis of surfactant free spherical nickel hydroxide/graphene oxide composite for supercapacitor application

A.A. Khaleed[1*], A. Bello[1], J. K. Dangbegnon[1], B. W. Mwakikunga[2], N. Manyala[1]

[1] Department of Physics, Institute of Applied Materials, SARCHI Chair in Carbon Technology and Materials, University of Pretoria, Pretoria 0028, South Africa
[2] DST/CSIR National Centre for Nano Structured Materials ,Council for Scientific and Industrial Research, P.O. Box 395, Pretoria 0001, South Africa
*abubakarkhaleed2@gmail.com

## INTRODUCTION

The growing need for energy coupled with the rise in global warming and air pollution due to the burning and consumption of fossil fuels has led to an increase in alternative energy sources to meet the exponential rise in energy demand [1]. These alternative energy sources often require a storage device as the energy production is not constant throughout the days and nights. Electrochemical capacitors (ECs) also known as supercapacitors (SCs) have attracted great attention in electrochemical energy storage application owing to their outstanding work rate, excellent cycle life and high power density [2]. In general, they are classified into two classes based on the mechanism in which they store charges: Electrochemical double-layer capacitors (EDLCs) which store energy based on ion adsorption, and faradaic capacitor which store energy by the fast surface redox process [3].

$Ni(OH)_2$, among many faradaic electrode materials has being considered as a potential electrode material due to its low cost, low toxicity, and high theoretical specific capacity (433 $mA{\cdot}hg^{-1}$) [4]. Nonetheless, $Ni(OH)_2$ suffers from low electrical conductivity, low rate capability and poor cycling stability, which affect its electrochemical performance and hence, restrict its commercialization [5]. Carbon nanomaterials such as graphene oxide (GO), carbon nanotubes (CNT) have been established to be effective in improving the electrochemical performance of $Ni(OH)_2$-based electrode materials when used as a supporting material [6].

## MATERIALS AND METHODS

### Synthesis of α-$Ni(OH)_2$/GO

2.33 g of $Ni(NO_3).6H_2O$ was dissolved in 40 mL of N, N-dimethylformamide (DMF) at room temperature under magnetic stirring, followed by the addition of 50 mg of GO and sonication for 1 h to get a uniform dispersion of GO in the solution. The solution was transferred into a 60 mL Teflon-lined autoclave and sealed. The autoclave was then placed into an electrical oven preheated at 180 °C and kept at that temperature for 18 h. Subsequently, the sample was removed from the oven and allowed to cool down to room temperature naturally. The obtained black powder was washed with deionized water until pH 7 was achieved and then dried at 60 °C overnight. Pure $Ni(OH)_2$ was synthesized via the same procedure with no addition of GO..

## RESULTS

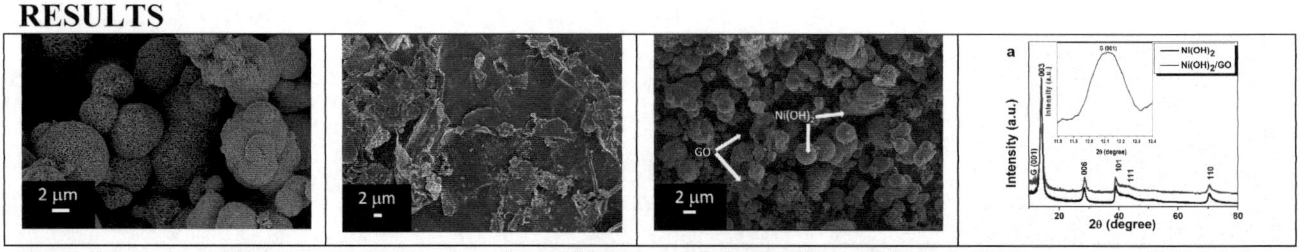


**Fig 1** High magnification SEM images of (a) $Ni(OH)_2$ and (b) $Ni(OH)_2$/GO; (c) XRD spectra of $Ni(OH)_2$ and $Ni(OH)_2$/GO samples

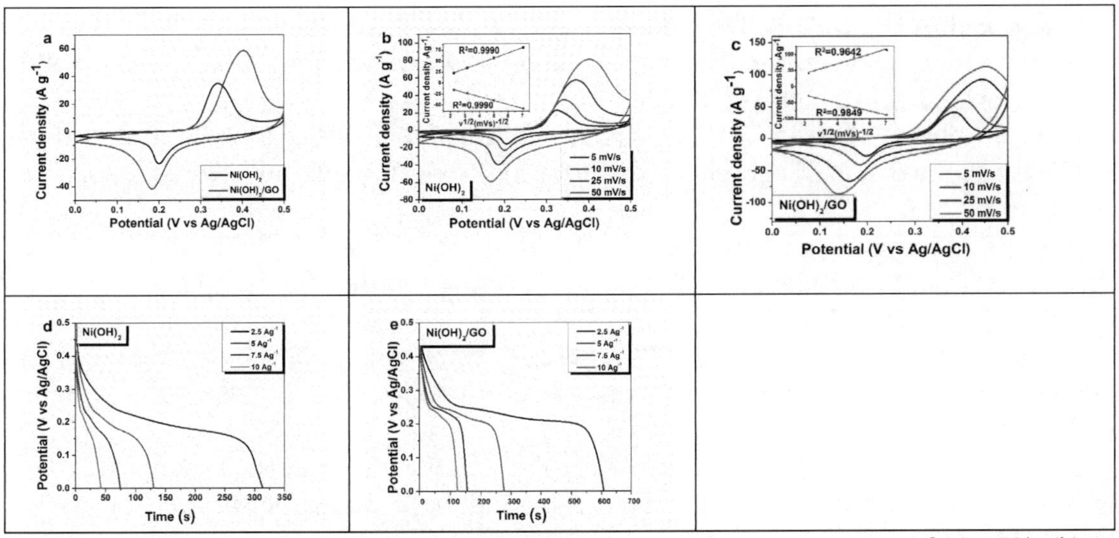


**Fig 2** (a) CV curves of $Ni(OH)_2$ and $Ni(OH)_2$/GF composite at sweep rate of 10 mV/s (b) and (C) CV curves of $Ni(OH)_2$ and $Ni(OH)_2$/GF composite at different sweep rates of 5-50 mV/s respectively; (d) and (e) CD curves of $Ni(OH)_2$ and $Ni(OH)_2$/GF composite at different current density ranging from 2.5-10 A/g

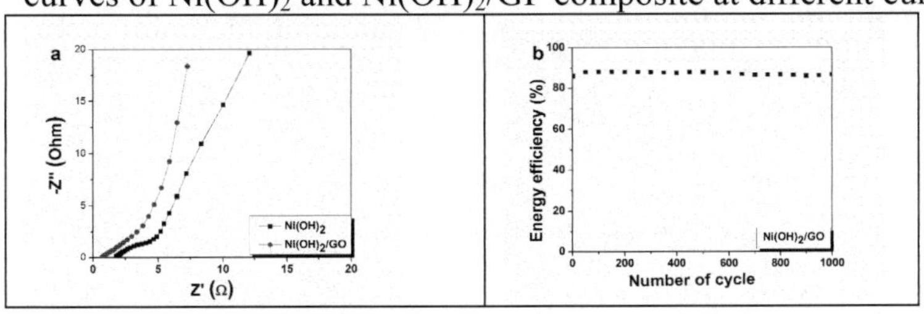


**Fig 4** (a) Nyquist plot of $Ni(OH)_2$ and $Ni(OH)_2$/GO electrodes (b) Cycling stability of the $Ni(OH)_2$/GF composite as a function of number of cycle at current density of 5 A/g.

## CONCLUSIONS

Spherical $Ni(OH)_2$/GO electrode was successfully synthesized using a surfactant free solvothermal procedure. The $Ni(OH)_2$/GO electrode displays better electrochemical performance as compared with pure $Ni(OH)_2$ electrode with a specific capacity of ~ 420 mAhg-1 which is close to the value of the theoretical capacity of $Ni(OH)_2$, excellent rate capability ~78 % at current density of 10 A g-1, with energy efficiency of ~87 % after 1000 at 5 $Ag^{-1}$. Hence, $Ni(OH)_2$/GO composite has a potential as an electrode material for ECs application.

## REFERENCES

[1] C.-Y. Cao, W. Guo, Z.-M. Cui, W.-G. Song, W. Cai, Microwave-assisted gas/liquid interfacial synthesis of flowerlike NiO hollow nanosphere precursors and their application as supercapacitor electrodes, J. Mater. Chem. 21 (2011)

[2] A.A. Khaleed, A. Bello, J.K. Dangbegnon, F.U. Ugbo, F. Barzegar, D.Y. Momodu, et al., A facile hydrothermal reflux synthesis of $Ni(OH)_2$/GF electrode for supercapacitor application, J. Mater. Sci. 51 (2016) 6041–6050.

[3] C. Jiang, B. Zhao, J. Cheng, J. Li, H. Zhang, Z. Tang, et al., Hydrothermal synthesis of $Ni(OH)_2$ nanoflakes on 3D graphene foam for high-performance supercapacitors, Electrochim. Acta. 173 (2015) 399–407.

[4] K. Wang, X. Zhang, X. Zhang, D. Chen, Q. Lin, A novel $Ni(OH)_2$/graphene nanosheets electrode with high capacitance and excellent cycling stability for pseudocapacitors, J. Power Sources. 333 (2016) 156–163.

[5] J.-H. Zhong, A.-L. Wang, G.-R. Li, J.-W. Wang, Y.-N. Ou, Y.-X. Tong, $Co_3O_4/Ni(OH)_2$ composite mesoporous nanosheet networks as a promising electrode for supercapacitor applications, J. Mater. Chem. 22 (2012) 5656.

[6] J. Huang, P. Xu, D. Cao, X. Zhou, S. Yang, Y. Li, et al., Asymmetric supercapacitors based on $\beta$-$Ni(OH)_2$ nanosheets and activated carbon with high energy density, J. Power Sources. 246 (2014) 371–376.

# Super-high Rate Sodium-Ion Storage Performance of $FeSe_2$/Graphene

Dan Li[1], Jisheng Zhou[1,2*], Xiaohong Chen[1], Huaihe Song[1,2*]

[1]State Key Laboratory of Chemical Resource Engineering, Beijing Key Laboratory of Electrochemical Process and Technology for Materials, Beijing University of Chemical Technology, Beijing 100029, P. R. China.
[2]Changzhou Institute of Advanced Materials, Beijing University of Chemical Technology, Jiangsu, P. R. China.
*zhoujs@mail.buct.edu.cn (Zhou, J), *songhh@mail.buct.edu.cn (Song, H)

## INTRODUCTION

Due to the scarcity of lithium resources, abundant, low-cost sodium-ion batteries (SIBs) have attracted more and more attention. However, the low specific capacity and dissatisfactory cycle performance make the sodium-ion battery difficult to realize practical implementation. Therefore, looking for high-performance anode materials is very important for SIBs. In this work, $FeSe_2$ nanorods uniformly anchored on the surfaces of graphene nanosheets ($FeSe_2$/GNSs) were successfully synthesized by a simple selenization process with Se powders. Our results showed that the $FeSe_2$/GNSs composite exhibits excellent sodium storage properties with a high capacity of 459 mAh/g and a remarkable rate capability simultaneously. The excellent electrochemical properties should be attributed to high purity of $FeSe_2$ as well as rod-like structure, which have higher conductivity and lower phase-transformation resistance, leading to more reversible conversion reaction. This method for design of iron selenide is also expected to be expanded for synthesis of other metal selenide composite materials, and improve their energy storage.

## MATERIALS AND METHODS

### Material

Iron(III) nitrate nonahydrate ($Fe(NO_3)_3 \cdot 9H_2O$, 98.5%) was purchased from Tianjin Guangfu Fine Chemical Industry Research Institute, China. Sodium borohydride ($NaBH_4$, 98%) was purchased from Tianjin Hainachuan Science and Technology Development Limited Company, China. Absolute alcohol (99.7%) was provided by Beijing Chemical Reagents Factory, China. The graphene nanosheets (GNSs) were fabricated by a modified Hummers method. Selenium powder was purchased from Alfa-Aesar Chemical Reagent Company. All of the chemicals used were AR grade. All the water used was deionized water ($18.2\Omega \cdot cm^{-1}$).

### Method

The nanorod $FeSe_2$/GNS composite was obtained through a simple selenization process with Se powders and iron oxide combined with graphene (referred to as $FeO_X$/GNS). The $FeO_X$/GNS composite was synthesized through an ice-bath reaction among $Fe(NO_3)_3$, graphene nanosheets and $NaBH_4$ solution for 4 h, as described in our previous work.[1] For the selenization process, $FeO_X$/GNS and excess selenium powders were loaded in a covered alumina boat and placed in a quartz-tube

reactor. Subsequently, the mixture was annealed at 400 °C for 12 h in atmosphere of a gas mixture

with 15% $H_2$/Ar to obtain $FeSe_2$/GNS composite.

## RESULTS AND DISCUSSION

All the XRD diffraction peaks (Figure 1a) in the $FeSe_2$/GNS composite match well with $FeSe_2$ without any impurity. The SEM image (Figure 1b) shows the curled morphology with a thin wrinkled microstructure. The TEM image (Figure 1c) shows that the $FeSe_2$ nanorods uniformly anchored onto the surfaces of GNSs.[2] A high capacitance of 459 mAh/g at the current density of 100 mA/g with excellent rate capacity was achieved by anchored $FeSe_2$ nanoparticles onto the surfaces of GNSs (Figure 1d). The reversible capacity at a high current of 5 A/g was remained at ca. 400 mAh/g after 800 cycles with no capacity decay. The excellent electrochemical properties should be attributed to the nanorod structure as well as strongly interfacial interaction[3] between $FeSe_2$ and GNS, which not only shortens the sodium ion diffusion path, but also benefits electronic transfer during the cycling.

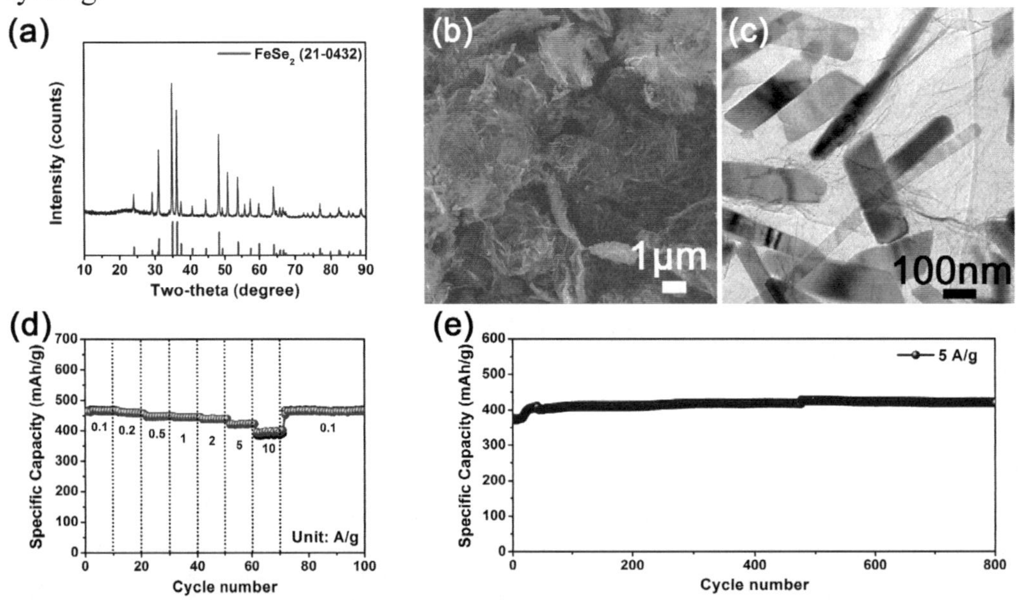


Figure 1. XRD pattern (a), SEM image (b) and TEM image (c) of as-prepared $FeSe_2$/GNS composites. Rate capability (d) and cycling performance at a high current of 5 A/g (f) for $FeSe_2$/GNS electrodes.

## CONCLUSIONS

In summary, we report a simple method of selenization to fabricate $FeSe_2$/GNS composites, and use it as anodes for sodium-ion batteries. The composite exhibits specific capacity as high as 459 mAh/g at 100 mA/g and good cycling properties, which should be due to the strongly interfacial interaction between $FeSe_2$ and GNS. This study clearly illustrates that $FeSe_2$/GNS with high- performance is a promising electrode material for sodium-ion batteries.

## REFERENCES


1. Li, D., Zhou, J., Chen, X. and Song, H. (2016). "Amorphous $Fe_2O_3$/graphene composite nanosheets with enhanced electrochemical performance for sodium-ion battery", *ACS Appl. Mater. Interfaces,* 8(45), 30899-30907

1. Zhang, K., Hu, Z., Liu, X., Tao, Z. and Chen, J. (2015). "$FeSe_2$ microspheres as a high-performance anode material for Na-ion batteries", *Adv. Mater.*, 27(21), 3305-3309

2. Zhou, J., Song, H., Ma, L. and Chen, X. (2011). "Magnetite/graphene nanosheet composites: interfacial interaction and its impact on the durable high-rate performance in lithium ion batteries", *RSC Adv.*, 1(5), 782-791

# Surface modification of carbon fibers by electroless tungsten deposition for enhanced electrochemical stability of Lead-Carbon interface

***Shu-Huei Hsieh*[1]**, Yi-Ren Tzeng[2], Ya-Wun Jan[3]

shhsieh@nfu.edu.tw

[1]Materials Science & Engineering, National Formosa University, Taiwan

[2]Division of Nuclear Fuels and Materials, Institute of Nuclear Energy Research, Taiwan

[3] Innolux Corporation, Taiwan

Energy storage plays a key role in the power management and broader use of renewable energy. The Electrochemical battery owns quick charge/discharge ability and high storage capacity is common used to store electric power. According to the report by the Sandia US national laboratory, a new battery design, known as the hybrid lead carbon battery formed by electronically connecting in parallel a traditional lead-acid battery and an asymmetric electrochemical supercapacitor, which provides a viable and economical method for energy storage. However, metals can't be well wetted on the carbon surfaces. Thus the metal contact of carbon electrode of hybrid lead carbon battery is usually assembled mechanically and gives rise to high contact resistance in use and eventually results in the battery will fail.

In this study, we present an easy and facile method to form electrochemically stable lead-carbon interface by using Tungsten carbides as a buffer layer to enhance the wetting of the lead on carbon materials. Luthin etc. found that carbon films on tungsten substrates could form small amount of WC at 870 k, therefore if Tungsten (W ) could be deposited on the surface of Carbon fibers, which will act as an anchor for catching Pb. In our experiments, the carbon fibers after Pb electroplating were modified by electroless W plating, and then were heated at vacuum to form chemical bonding at lead-carbon interface.

The experimental results reveal there are many Pb particles with 100~200 nm size uniformly dispersed on the surface of carbon fibers. Fig. 1 and 2 are the SEM photos and SEM/EDS analysis of Pb/W/Carbon, separately. The SEM morphology shows that most Pb particles were tiny and did not agglomerate on the surface of carbon fibers, the results confirm that Pb-C interface has good adhesion. The chemical composition of Pb/W/Carbon Fibers lists at Fig. 2, including Pb and W, the others like Sn, P and Na came from electroless plating, C and O should be Carbon fiber and Oxidation of carbon or lead.

**Keywords：electroless W plating, Carbon fiber, lead-carbon (Pb-C) interface**

Fig 1 SEM photos of Pb/W/Carbon Fiber composite material after vacuum heat treatment at 600 □.

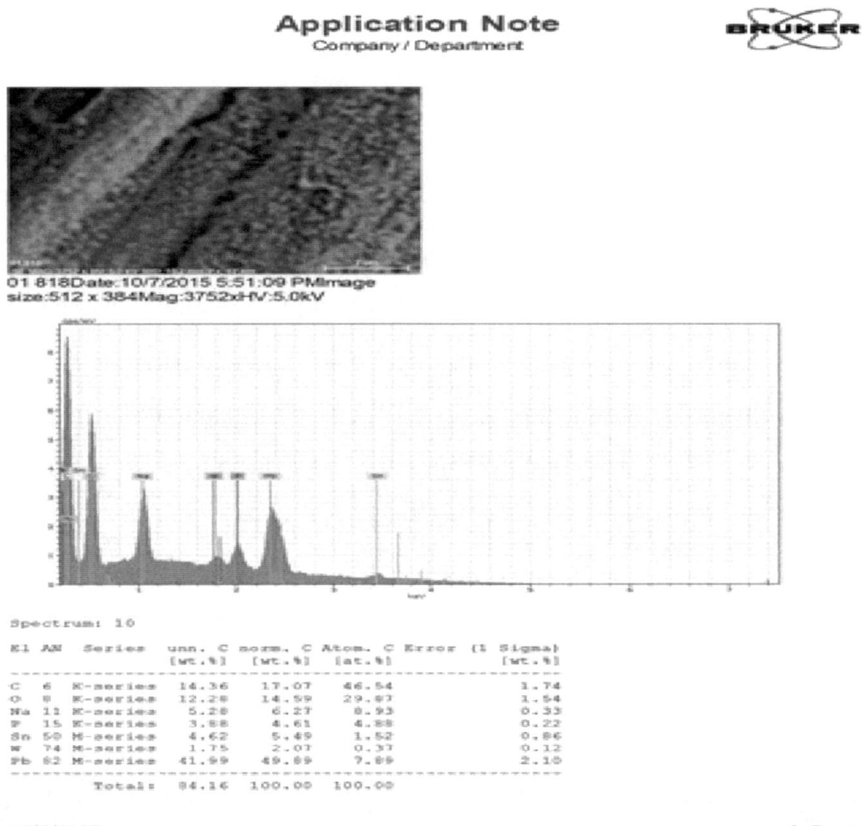
Application Note
Company / Department
BRUKER

01 818Date:10/7/2015 5:51:09 PMImage size:512 x 384Mag:3752xHV:5.0kV

Spectrum: 10

| El | AN | Series | unn. C [wt.%] | norm. C [wt.%] | Atom. C [at.%] | Error (1 Sigma) [wt.%] |
|---|---|---|---|---|---|---|
| C | 6 | K-series | 14.36 | 17.07 | 46.54 | 1.74 |
| O | 8 | K-series | 12.28 | 14.59 | 29.87 | 1.54 |
| Na | 11 | K-series | 5.28 | 6.27 | 8.93 | 0.33 |
| P | 15 | K-series | 3.88 | 4.61 | 4.88 | 0.22 |
| Sn | 50 | M-series | 4.62 | 5.49 | 1.52 | 0.86 |
| W | 74 | M-series | 1.75 | 2.07 | 0.37 | 0.12 |
| Pb | 82 | M-series | 41.99 | 49.89 | 7.89 | 2.10 |
| | | Total: | 84.16 | 100.00 | 100.00 | |

10/7/2015 1. Page

Pb

W

Fig 2 SEM/EDS analysis of Pb/W/Carbon Fiber composite material after vacuum heat treatment at 600 □.

# SURFACE MODIFICATION OF GRAPHENE AND GRAPHENE OXIDES BY NITROGEN PLASMA: DETERMINATION OF CHEMICAL STATE ALTERATIONS BY QUANTITATIVE X-RAY PHOTOELECTRON SPECTROSCOPY

Krisztina László[1*], Barbara Berke[1,2], Imre Bertóti[3], Miklós Mohai[3]

[1] Budapest University of Technology and Economics, Budapest, Hungary
[2] Institute Laue Langevin, Grenoble, France
[3] Institute of Materials and Environmental Chemistry, Research Centre for Natural Sciences, Hungarian Academy of Sciences, Budapest, Hungary
*klaszlo@mail.bme.hu

## INTRODUCTION

The various applications of nanocarbon materials including graphene and graphene oxides often require surface modification. Radio frequency (RF) cold plasma activation is a simple method for initiating gas-solid reactions to modify the surface. This method enables one to implant the plasma ions into the sub-surface atomic layers, accommodating them in different covalent bonding states.
In this work thin films of graphene, graphene oxide and reduced graphene oxide were treated by low pressure RF $N_2$ gas plasma, nominally at room temperature.

## MATERIALS AND METHODS

Multi-layer graphene powder (GR), obtained by mechanical exfoliation of high purity synthetic graphite (Aldrich, USA) and single layer graphene oxide (SLGO) (Cheap Tubes Inc, USA) were used as model materials. Plasma treatment was performed in the sample preparation chamber of the XPS instrument (base pressure $<1\cdot10^{-4}$ Pa), enabling in situ analysis. The high purity (5N) $N_2$ flow was regulated to set working pressure to $3\cdot10^{-1}$ Pa. RF power of 100 W at 13.56 MHz was applied. The sample bias was set to 0 V (earthed) or to negative values of 50-300 V, varying the ion energy of N ions between 0 and 150 eV. A fixed 10 min treatment time was selected.
X-ray photoelectron spectra were recorded on a Kratos XSAM 800 spectrometer. Intensity data were obtained after Shirley type background removal. Quantitative analysis was performed by the XPS MultiQuant 7.6 program using the experimentally determined photo-ionisation cross-section data of Evans *et al.*

## RESULTS AND DISCUSSION

At identical conditions (200 V bias) similar amount, 9.3 atomic % N was implanted in graphene and SLGO. Significant reduction of SLGO took place at the same treatment: the initial 28 atomic % O was decreased to 23%. Three different chemical bonding states of the nitrogen were found (see **Fig.**) for all treated samples and assigned to pyridine-type (398.3 eV), pyrrole-type ring structures (399.7 eV), and substitutional N in a graphite-like environment (400.9 eV). While in the GR the pyridine, in SLGO the pyrrol type states are prevailing. For the carbon, the C1s spectral regions are depicted before and after the plasma treatment (see **Fig.**). Six different chemical states of carbon were identified (identical for both treated and untreated samples): $sp^2$ C-C in graphene plane (284.3 eV); $sp^3$ C-C (285.6 e); C in hydroxyl or C–N bond in pyridine ring or amine (285.8 eV); C in ether or epoxy bond (286.6 eV); C in carbonyl bond, C–N bond in graphene plane (287.6 eV); C in carboxyl or ester bond (288.5 eV).

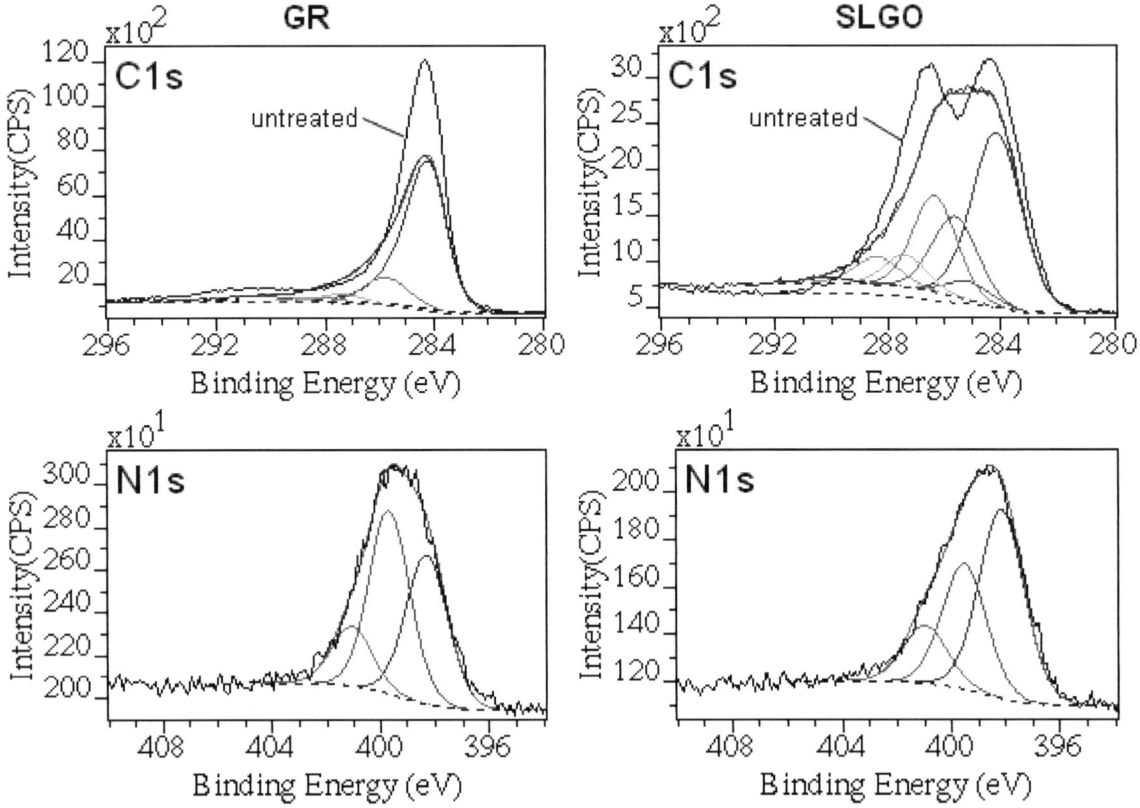


Experiments, performed using similar conditions on other nanocarbon materials (e.g., nanotubes or graphite), showed that similar chemical states were developed, although with different quantities.

**CONCLUSIONS**

It is demonstrated that the surface of GR and SLGO, can be conveniently modified using RF plasma activated low pressure nitrogen. The amount of nitrogen can be tuned by varying the negative sample bias. The implanted nitrogen is incorporated in the top 1-6 atomic monolayers of the samples.

In case of SLGO carbonyl (C=O) and carboxyl (O–C=O) bonding states were found to be more reactive, relative to others, showing significant decrease at all plasma treatments.

**ACKNOWLEDGEMENTS**

Sponsored by the National Scientific Research Fund (OTKA) project No. K109558. The work is part of the EU project FP7-PEOPLE-2010-IRSES-269267 and the Nanoporous and Nanostructured Materials for Medical Applications (Nanomed: H2020-MSCA-RISE-2016, project:734641).

**REFERENCES**

1. Bertóti, I., Mohai, M. and László, K. (2015). "Surface modification of graphene and graphite by nitrogen plasma: Determination of chemical state alterations and assignments by quantitative X-ray photoelectron spectroscopy", *Carbon*, Vol.84, p.185-196.

2. Bertóti, I. and Mohai, M. (2016). "Modification of graphene-oxide surface in nitrogen and argon glow discharge plasma", *Surf. Interface Anal.* Vol.48, p.461-464.

3. Bertóti, I., Mohai, I., Mohai, M. and Szépvölgyi, J. (2011). "Surface modification of multi-wall carbon nanotubes by nitrogen attachment", Diamond Relat. Mater., Vol.20, p.965-968.

# Synthesis of frost-like CuO combined graphene-$TiO_2$ for high photocatalytic performance

Dinh Cung Tien Nguyen [1], Woo-Sik Kim [2], Kwang-Youn Cho [2], and **Won-Chun Oh** [1 *]

(E-mail: wc_oh@hanseo.ac.kr)

[1] Department of Advanced Materials Science & Engineering, Hanseo University, Seosan, Chungnam, Korea, 356-706

[2]Korea Institutes of Ceramic Engineering and Technology, Soho-ro, Jinju-Si, Gyeongsangnam-do, Republic of Korea

## INTRODUCTION

$TiO_2$ materials only show the catalytic effect of the ultraviolet light (UV) because of the large band gap energy, about 3.2 – 3.5 eV. CuO also shows some advantageous properties such as high theoretical capacity, high safety, environmental benignity, and low cost [1, 2]. Besides, the combination of CuO and $TiO_2$ promises to be feasible, novel, and environmentally friendly, while overcoming the drawback of $TiO_2$ while contributing to the creation of a new semiconductor material. Since CuO nanostructures have a large surface area, the specific capacitance of this material has been improved, notably its stability [3]. In this study, a frost-like CuO combined graphene-$TiO_2$ composite using a self-assembly method was prepared. Specifically, Cu salt was dissolved in distilled water, and the obtained solution was stirred at a medium temperature after mixing with graphene oxide and a $TiO_2$ precursor. The structure and morphology of the frost-like CuO combined graphene-$TiO_2$ composite were characterized by XRD, SEM, EDX, TEM, Raman spectroscopy, UV-vis-DRS, and XPS. Photodegradation experiments were also conducted with organic dyes such as rhodamine B (RhB), methylene blue trihydrate (MB), and reactive black B (RBB) in an aqueous solution under visible light irradiation. To elucidate the reaction mechanism and investigate the specific reactive species involved in the photodegradation, the scavenging experiments were also carried out. The structural stability of the photocatalytic materials was presented to provide evidence to confirm the stability after recycling measurements.

## METHODOLOGY

The Hummers' Method was applied for the preparation of graphene oxide [4]. 0.005 mol $Cu(CH_3COO)_2.H_2O$ was dissolved in 50 mL distilled water, then heated to 60°C with magnetic stirring for 30 min. Meanwhile, the graphene oxide (0.01 g) was sonicated in 10 ml ethanol 95% for 30 min, followed the addition of $TiO_2$ nanopowder (0.1 g) and PVP (0.1 g) to obtain the CuO-graphene-$TiO_2$ composite. Eventually, for each composite synthesis process, the pH value was adjusted with $NH_4OH$ 25% until reaching pH 9 and was then maintained for 3 hours. The product was washed with distilled water 3 times and 95% ethanol twice. After drying under vacuum at 105°C for 24h.

## RESULTS AND DISCUSSION

The patterns in the XRD shows the three types of main diffraction peaks of the CuO-graphene-$TiO_2$ composite with graphene, CuO, and $TiO_2$ signals. The surface morphology of the CuO-graphene-$TiO_2$ was investigated by SEM and TEM. The band gap energy values of CuO-graphene-$TiO_2$ correspond to 2.79 eV which is lower than that of the CuO-$TiO_2$ composite. From the XPS results, the surface bonding state of the survey nanomaterial was identified as containing Cu 2p, Ti 2p, O 1s, and C 1s, reflecting the elemental composition of the nanomaterial without the presence of another signal. Through the Raman results, the formation of new links in the CuO-graphene-$TiO_2$ nanocomposite can be demonstrated through a blue-shift of the CuO signal compared to the peak of CuO in the CuO-graphene (278 and 620 $cm^{-1}$) and CuO-graphene-$TiO_2$ (270 and 614 $cm^{-1}$) composites. Oversing the photodegradation results, the degradation ratio (Ct/Co) of CuO-graphene-$TiO_2$ is much higher than that of the other catalysts used to produce dyes of RhB, MB, and RBB. The scavenging experiments confirmed that $^{\bullet}OH$ and $h^+$ play a major role in the photocatalytic reaction than $O_2^{\bullet-}$ or both hydroxyl and holes are the active species responsible for the RBB degradation under visible light irradiation. Besides, the photocatalytic activity of CuO-graphene-$TiO_2$ was exhibited good result after five recycling runs.

## CONCLUSION

In this study, we successfully synthesized CuO-graphene-$TiO_2$ composite by the self-assembly method. These results revealed that CuO-graphene-$TiO_2$ is a new potential material for high photodegradation activity.

**References:**


[1] Bao Wang, Xing-Long Wu, Chun-Ying Shu, Yu-Guo Guo and Chun-Ru Wang, Synthesis of CuO/graphene nanocomposite as a high-performance anode material for lithium-ion batteries, J. Mater. Chem, 20 (2010) 10661–10664.

**[2] X. Zhang, W. Shi, J. Zhu, D. J. Kharistal, W. Zhao, B. S. Lalia, H. H. Hng and Q. Yan, High-power and high-energy-density flexible pseudocapacitor electrodes made from porous CuO nanobelts and single-walled carbon nanotubes, ACS Nano, 5 (2011) 2013-9. Doi: 10.1021/nn1030719**

**[3] C. D. Lokhande, D. P. Dubal and O. S. Joo, Metal oxide thin film based supercapacitors, Curr. Appl. Phys., 11 ( 2011) 255-270.**

[4] R. Sahraei, Z. S. Pour, M. Ghaemy, Novel magnetic bio-sorbent hydrogel beads based on modified gum tragacanth/graphene oxide: Removal of heavy metals and dyes from water. Journal of Cleaner Production, 142 (2017) 2973-2984.

# Synthesis of graphene based materials and their fuel cells electrocatalyst applications

María Jesús Lázaro[1*], Luis M. Rivera[2], César Montero[2], Gonzalo García[2] and Elena Pastor [2*]

[1] Instituto de carboquímica (CSIC), Miguel Luesma Castan 4, 50018 Zaragoza, Spain
[2] Instituto de Materiales y Nanotecnología, Universidad de La Laguna, Apartado Postal 456, 38200 La Laguna, Santa Cruz de Tenerife, Spain
*mlazaro@icb.csic.es, epastor@ull.edu.es

## INTRODUCTION

The worldwide energy demand continuously increases and the search for novel energy conversion process becomes necessary. In this context, Fuel Cells (FCs) emerge as attractive alternative for clean energy production and graphenic materials (GMs) appear as promising catalyst supports due to their excellent conductivity and high surface area. In order to produce GMs, graphene oxide (GO) is commonly used as low-cost precursor. However, GO presents low electric conductivity due to the presence of oxygenated groups formed during the synthesis from graphite, which restricts its electrocatalytic applications. To solve this problem a great diversity of physical and chemical reduction methods were reported in the literature[1]. Chemical reduction is a cheap and versatile method to produce reduced graphene oxide (RGO) doped with P, B, S and N elements[2]. Doping process causes an electronic modulation of the carbonaceous grid improving the catalytic activity toward several reactions of interest. For example, it was reported higher performances of Co-doped RGO than N-doped RGO as cathodic material for FCs[2].

In the present work, several GMs have been synthesized using a chemical reduction method employing ethylene glycol (EG-rGO), hydrazine (H-rGO) and thiourea (SN-rGO) as reducing agents. Then, these materials were used as supports for Pd and Ni nanoparticles. Finally, performances towards the oxygen reduction (ORR) and hydrogen evolution (HER) reactions were evaluated in basic media.

## MATERIALS AND METHODS

GO was synthesized using the modified Hummers method[3]. Then H-rGO was prepared refluxing a GO dispersion with appropriate amount of hydrazine during 12 h; EG-rGO by dispersion of GO in ethylene glycol during 12 h in magnetic stirring; and SN-rGO by an hydrothermal method, mixing

GO and thiourea and placing the mixture into a Teflon vessel for digestion during 10 h at 160°C in

an oven.

Carbon domains and crystallinity were determined using Raman spectroscopy and X-ray diffraction (XRD). Morphology and elemental composition were established from scanning electron microscopy (SEM) and X-ray dispersive energy (EDX).

Cyclic voltammetry (CV) and linear sweep voltammetry (LSV) using a rotating disk electrode (RDE) were applied. All experiments were carried out in 0.1 M NaOH as supporting electrolyte. ORR and HER polarization curves were measured at 2 and 1 $mVs^{-1}$, respectively.

## RESULTS AND DISCUSSION

Figure 1a shows Raman profiles for GMs. Several changes produced by the chemical treatments can be seen. Thus, $I_D/I_G$ intensity peak ratios for rGOs are < 1, except for SN-rGO ($I_D/I_G$ =1.19). This value can be justified in terms of the presence of $sp^3$ carbon hybridization related to S and N heteroatoms in the stucture[2]. On the other hand, interplanar spacing increases after graphite oxidation to GO ($d_{002}$ = 0.84 nm) and recovers the former value (around 0.4 nm) after reduction treatments (Figure 1b).

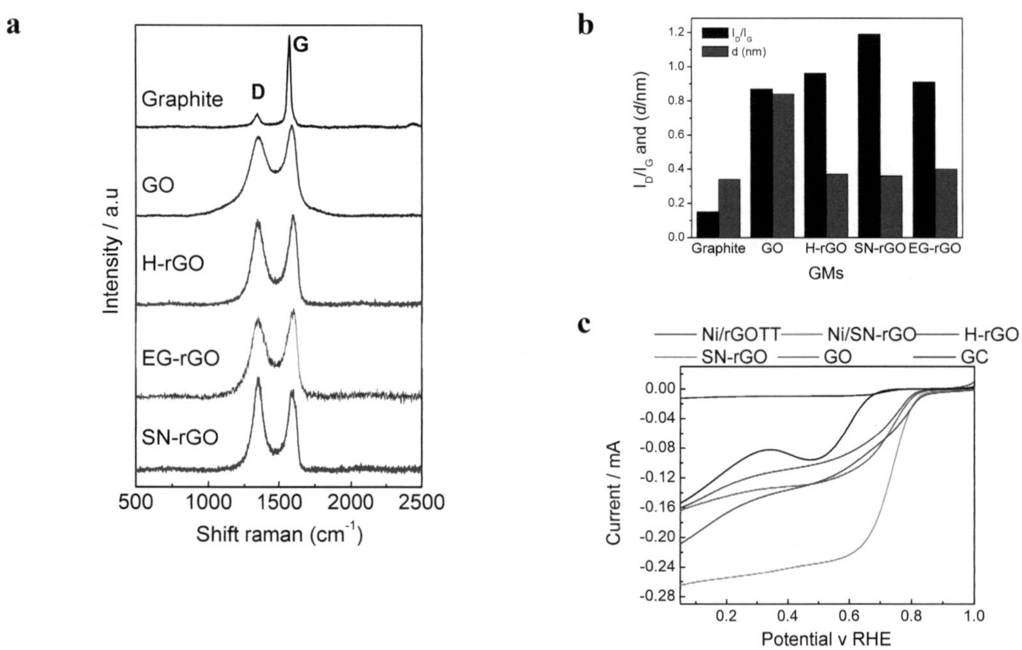


**Figure 1.** a) Raman spectra of synthesized GMs . b) Interplanar spacing and $I_D/I_G$ ratio for the GMs. c) ORR polarization curves.

Moreover, the doping process induces changes in the ORR and HER responses. Figure 1c displays the ORR curves for Ni catalysts. An important increment of the catalytic activity is observed when S and N heteroatoms are inserted into the graphene network. On the other hand, GO presents a poor ORR performance compared to the rest of GMs due to its low conductivity. Also doping modifies HER performances (not shown).

## CONCLUSIONS

GMs have been successfully synthesized using different reducing agents and chemical procedures. Spectroscopic and X-ray techniques allowed to establish the changes in the physicochemical properties produced by the presence of the heteroatoms introduced during the reduction process. Accordingly, GMs present diverse ORR and HER performances.

## ACKNOWLEDGEMENTS

This work has been supported by the Spanish Ministry of Economy and Competitiveness under project ENE2014-52158-C2-2R (co-funded by FEDER). LMR and CM acknowledge the Canary Government (ACIISI) and MINCEO for their pre-doctoral grants respectively. GG acknowledges the Viera y Clavijo program (ACIISI & ULL) for financial support.

**REFERENCES**

1. Pei, S., Cheng, H-M. (2011) "The reduction of graphene oxide", Carbon, Vol. 50(9), p. 3210–3228

2. Su, Y., Zhang, Y., Zhuang, X., Li, S., Wu, D., Zhang, F., Feng, X. (2013) "Low-temperature synthesis of nitrogen/sulfur co-doped three-dimensional graphene frameworks as efficient metal-free electrocatalyst for oxygen reduction reaction", Carbon, Vol. 62, p.296-301.

3. Flórez-Montaño, J., Calderón-Cárdenas, A., Lizcano-Valbuena, W., Rodríguez, J.L., Pastor, E. (2016) "Ni@Pt nanodisks with low Pt content supported on reduced graphene oxide for methanol electrooxidation in alkaline media", Vol. 41(43), p.19799–19809.

## Tailoring Few-Layer Graphene for Supercapacitors Using a Hydrogen-Argon Mixed Plasma

Emily A M Smith[1 2*], Carol Crean[1], Robert Slade[1], Chris Spacie[2], Chris Stirling[2], David Watson[1]

[1]Department of Chemistry, University of Surrey, Guildford, Surrey UK
[2]Haydale Ltd, Ammanford, Carmarthenshire UK.
*e.a.smith@surrey.ac.uk

### INTRODUCTION

Haydale Ltd have a patented, large scale process for plasma functionalisation of 2-dimensional nanomaterials.[1] Cold, low pressure (< 100 Pa) and low frequency (< 100 kHz) plasma can be used to uniformly treat carbon nanomaterial powders and add purpose-specific chemical groups onto the surface, which allow the carbon nanomaterials to disperse into any chosen matrix and improve their performance in electrochemical applications with pseudo-capacitive enhancements. The focus of this work is use of differing functionalisation processes on few-layer graphenes (FLG). The application of the such functionalised FLG materials in aqueous supercapacitors has been investigated.

### MATERIALS AND METHODS

Commercially available FLGs were processed in a Haydale HT-60 R&D reactor in different process gases sourced from the BOC Group. Argon gas and a 3% hydrogen in argon gas mixture was used. Scanning electron microscopy (SEM) images were obtained with a JEOL JSM-7100F model. The X-ray photoelectron spectroscopy (XPS) data were acquired from a Theta Probe Angle-Resolved X-ray Photoelectron Spectrometer (ARXPS) system; the instrument used monochromatic Al $K_\alpha$ radiation (1486.6 eV) with an X-ray spot size of 400 μm. The spectra were processed using CasaXPS software with Shirley backgrounds. The C 1s and O 1s peaks were fitted with consistent components across all materials using relevant literature as comparison.[2,3] The carbon pellet electrodes consisted of 90% dried FLG, 5% polytetrafluoroethylene (PTFE) binder and 5% Super P carbon black. The current collectors were stainless steel. Two such electrodes were separated by glassy fibrous paper in a two-electrode sandwich-type capacitor with aqueous sulphuric acid (2M) electrolyte. Cyclic voltammetry (CV) was carried out at scan rates 1 - 50 mV $s^{-1}$ on an Autolab-Ecochimie PGSTAT 30. The potential window was 0 - 0.8 V. The supercapacitor testing results presented were carried out by a third party, INCAR/CSIC, and reported to Haydale Ltd.

### RESULTS AND DISCUSSION

The FLGs were placed in an oven for 18 hours at 130°C (FLG-Dried). They have platelet morphology, which agglomerates into seemingly larger particles. The primary particle sizes are less than 100 nm to 1 μm in lateral dimension and less than 2 nm in thickness (2 - 12 graphene layers).

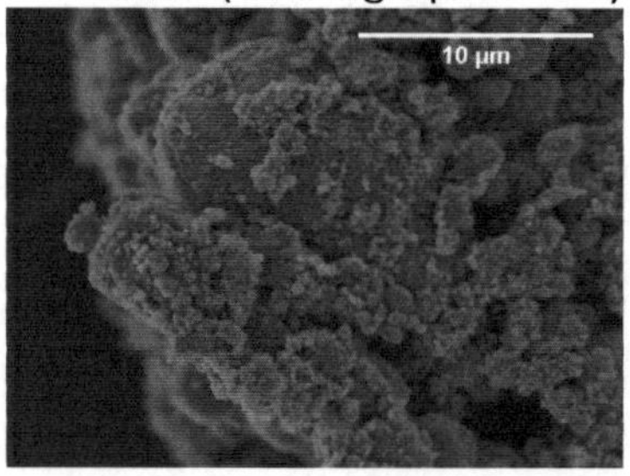


Fig. 2. SEM image of FLG-Dried.

The FLG-Dried was then processed in argon (Ar) or hydrogen-argon mixed (H/Ar) plasmas. The XPS results were used to analyse the changes in the surface nature of the materials. The process in general introduces oxygen into the material, but there are differing quantities of oxygen-functional groups. The starting material had a higher concentration of alcohol, ether, epoxy and ester groups (~ 532.2 eV) and a lower concentration of carbonyl (~530.8 eV) and carboxyl (~533.3 eV) groups. The Ar-treated material increases the amount of carbonyl, alcohol, ether, epoxy and ester groups to a larger extent than the H/Ar process.

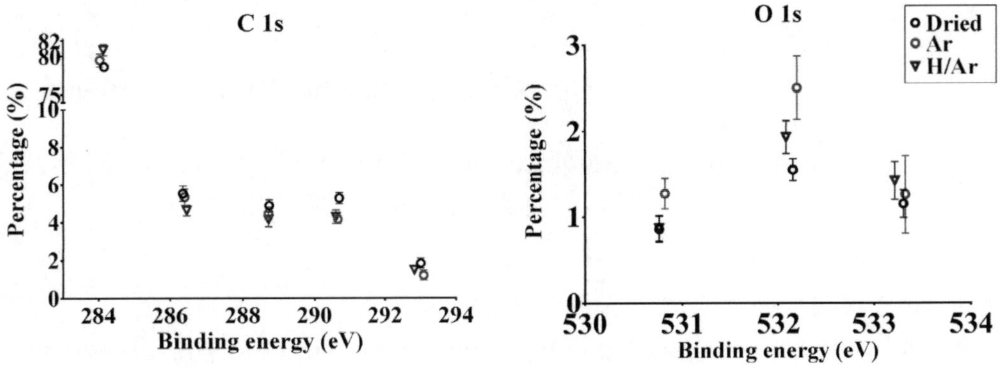


Fig. 3. XPS C 1s and O 1s peak components: the average percentages and standard deviations (n = 9).

When cycled in supercapacitor test devices, the H/Ar material showed an improved capacitance compared to the Ar material.

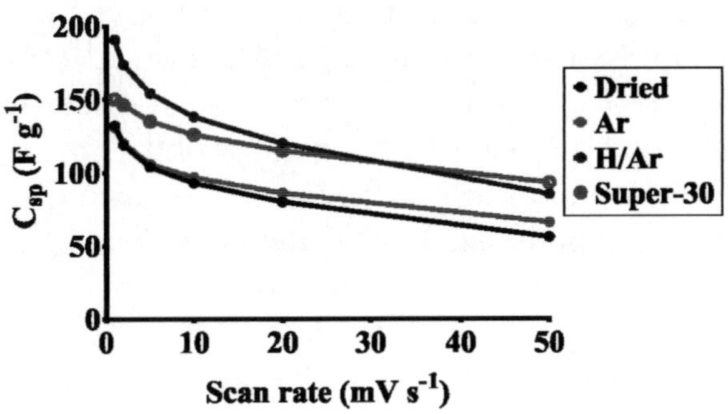


Fig. 3. Specific capacitance values from cyclic voltammetry testing of supercapacitors at scan rates of 1 - 50 mV s$^{-1}$. Super-30 is a commercial activated carbon with a surface area of 1100 m$^2$ g$^{-1}$, much higher than that for the FLG materials.

## CONCLUSIONS

The original pseudo-capacitance for FLG devices provided by existing oxygen surface groups, undergoes different enhancements after plasma treatment. The H/Ar mixed plasma significantly improves the performance in supercapacitor electrodes.

## REFERENCES

1. Walters, I. and Williams, M. (2012). "Particulate materials, composites comprising them, preparation and uses thereof", WO2012076853 A1

2. Bertóti, I., Mohai, M. and László, K. (2014). "Surface modification of graphene and graphite by nitrogen plasma: determination of chemical state alterations and assignments by quantitative X-ray photoelectron spectroscopy", *Carbon*, Vol.84, p.185-196

3. Lin, Y. C., Li, C. Y. and Chiu, P. W. (2010). "Controllable graphene N-doping with ammonia plasma", *Appl. Phys. Lett.*, Vol.96, p.13310

# Thermodynamic rationalization of graphitization: the critical temperature threshold

Philippe Ouzilleau[1], Aïmen E. Gheribi[1], Patrice Chartrand[1*]

[1]Department of Chemical Engineering, Polytechnique Montreal, Montreal, Canada
*philippe.ouzilleau@polymtl.ca

## INTRODUCTION

Previous application of kinetic equations to model the effect of heat treatment temperature on the graphitization behaviour of carbon materials is well known. However, these studies could not explain apparently controversial experimental results, hindering conceptualization of key aspects of the graphitization mechanism.

The present work focuses on developing a thermodynamic rationalization of the irreversible graphitization process based on the concept of a critical temperature ($T_c$) at ~2550 K [1]. This thermodynamic model reflects how the concept of a critical temperature threshold for the surface energy of coke crystallites [2] (idealized, non-graphitic, graphitizable crystallites) constitutes the first step of the presently proposed graphitization scenario. The model is consistent with available graphitization experimental knowledge. This includes the experimental temperature threshold for the onset of true long range graphitic order, the reported apparent formation of linear carbon chains in graphite and the Oberlin [3] polygonization temperature, all three phenomena apparently linked to a critical temperature of ~2550 K.

Quantitative results for the validation of the model are given. Predictions for the average interlayer spacings ($d_{002}$), the thermal diffusivities and the electrical resistivities as a function of graphitization temperature (between 2000 and 3300 K) of 51 graphitizable carbons (petroleum coke, polyvinyl chloride, anthracites, mesocarbon microbeads, polyimide films, etc.) have been presented in details [1]. Predictive accuracy is statistically evaluated. Near-ideal precision is obtained.

## RESULTS AND DISCUSSION

The presently proposed rationalization of graphitization is presented in Fig. 1. This scenario assumes that coke crystallites are independent entities below the critical temperature threshold of 2550 K and that this independence is lost at the critical temperature. The driving force for the progressive merging and flattening of the coke crystallites below 2550 K is thus the progressive removal of cross-linking intercrystalline material (white circles in Fig. 1) as this material is the key factor for maintaining relative crystallite independence.

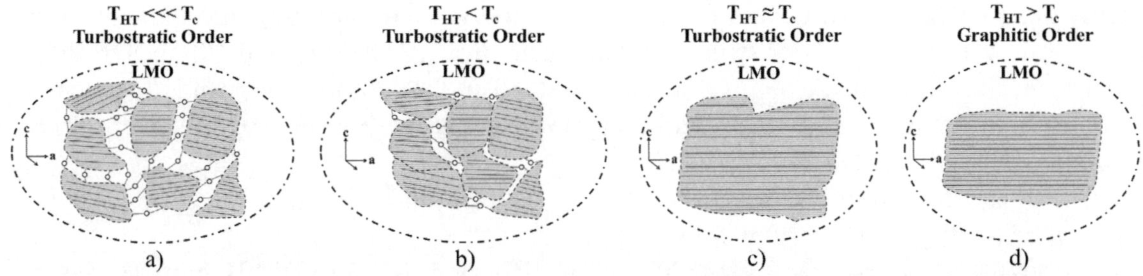


**Figure 7: Rationalization of graphitization to model the structural effect of the heat treatment temperature ($T_{HT}$) near the critical temperature ($T_c$) of 2550 K**

By modelling the graphitization process in such a way as shown in Fig. 1, a methodology similar to other thermodynamic Second-Order Structural Transformations can be applied to model the process [1]. It is possible to produce thermodynamic predictions for quantitative values of key graphitization properties such as the $d_{002}$ (Fig. 2) as a function of the heat treatment temperature ($T_{HT}$).

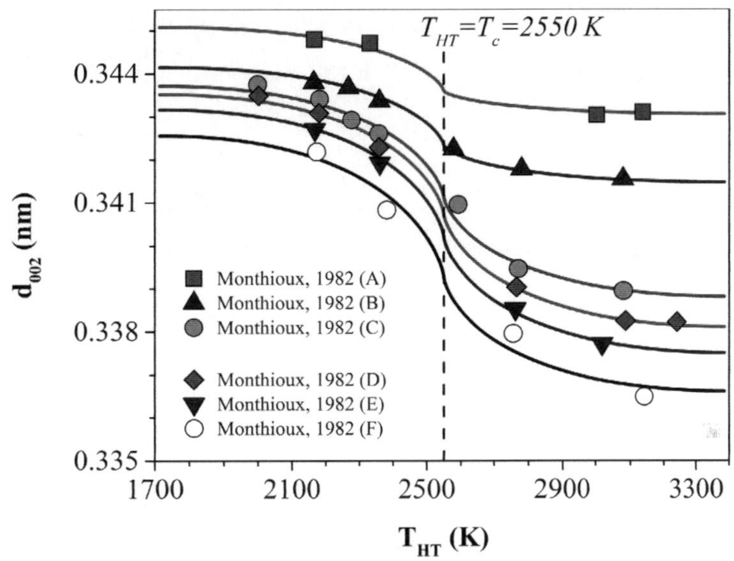


**Figure 8: Predictions of $d_{002}$ as a function of $T_{HT}$ based on the present Second-Order Structural Transformation model of the graphitization process of Fig. 1. Experimental measurements taken from Monthioux[4].**

## CONCLUSIONS

The present graphitization rationalization provides a thermodynamic basis for future models focused on the graphitization mechanism. Some preliminary comments on the modelling of the graphitizable/non-graphitizable character of various carbons, based on the presently discussed thermodynamic principles, will be discussed in the presentation. Key conclusions of the present work are:

-Long-range graphitic order develops if $T_{HT} > T_c$ in graphitizable carbons heat treated at normal pressures.

-The complete removal of intercrystalline matter explains the divergence in $d_{002}(T_{HT})$ at $T_{HT} = T_c$ shown in Fig. 2 .

-The Oberlin polygonization temperature is consistent with the proposed existence of $T_c$

## REFERENCES

1. Ouzilleau, P., Gheribi, A. E. and Chartrand, P. (2016). "The graphitization temperature threshold analyzed through a second-order structural transformation", *Carbon.*, Vol.109, p.896-908
2. Ouzilleau, P., Gheribi, A. E., G. Eriksson, D. Lindberg and Chartrand, P. (2015). "A size-dependent thermodynamic model for coke crystallites: the carbon-hydrogen system up to 2500 K", *Carbon.*, Vol.85, p.99-118
3. Oberlin, A., Bonnamy, S. and Oshida, K. (2006). "Landmarks for graphitization", *Tanso*, Vol.224, p.281-298
4. Monthioux, M., Oberlin, M., Oberlin, A. and Bourrat, X. (1982). "Heavy petroleum products: microtexture and ability to graphitize", *Carbon*, Vol.20, p.167-176

# Thermophysical properties of carbon fibre reinforced multilayered (PyC–SiC)$_n$ matrix composites

Yan Jia[1*], Kezhi Li[1]

[1]Northwestern Polytechnical University, Xi'an, P.R. China
*jiayannwpu@163.com

## INTRODUCTION

Carbon fibre reinforced carbon-silicon carbide matrix (C/C–SiC) composites have attracted considerable attention as thermo-structural materials in aeronautical and astronautical fields [1]. These strategic applications require C/C–SiC composites to have low and uniform coefficient of thermal expansion (CTE) and high thermal conductivity in addition to high specific strength and fracture toughness [2]. Therefore, it is essential to investigate the thermophysical properties of C/C–SiC composites as important design factors.

In our previous work [3], a new member of C/C–SiC composites was fabricated by alternate deposition of pyrocarbon (PyC) and SiC inside preforms via chemical vapour infiltration (CVI), i.e. carbon fibre reinforced multilayered (PyC–SiC)$_n$ matrix (C/(PyC–SiC)$_n$) composites. In addition, mechanical properties of the composites were studied intensively. In this work, thermal expansion behaviours and thermal conductivity of the composites were investigated. The relationships between the thermophysical properties and matrix microstructures were also discussed in detail.

## MATERIALS AND METHODS

### Materials

2.5D needle-punched carbon fibre felts with a density of 0.4-0.45 g/cm$^3$ were used as preforms. CVI was employed to alternately infiltrate PyC and SiC into preforms. One PyC–SiC sequence, two PyC–SiC sequences and four PyC–SiC sequences were deposited on carbon fibres to prepare three kinds of C/(PyC–SiC)$_n$ composites (designated as CS1, CS2 and CS4, respectively) The final bulk density and open porosity of each composite were 1.34 g/cm$^3$ and 35.1%, 1.33 g/cm$^3$ and 37.2%, 1.36 g/cm$^3$ and 34.6%, respectively.

### Methods

Cylinder samples with dimensions of Φ6 mm × 10 mm were used for CTE measurement performed on a DIL 402E dilatometer (NETZSCH, Germany). All measurements were conducted in a high-purity helium atmosphere from 800 to 2500 ºC. The testing direction was perpendicular (Z direction) to the non-woven cloth. Thermal diffusivity tests were measured on a laser flash analyser (LFA 427, NETZSCH, Germany). The samples were machined into discs with dimension of Φ12.7 mm × 3 mm and tested in Z direction from RT to 1900 ºC. Pore size distribution of the composites was determined by a mercury porosimeter (Poremaster 33, Quantachrome, USA). Microstructure of the multilayered matrix was observed by scanning electron microscope (SEM, VEGA3, Tescan, Czech) equipped with an energy dispersive spectroscope (EDS, Oxford INCA, UK).

## RESULTS AND DISCUSSION

Fig. 1 shows the backscattered electron (BSE) images of polished cross-sectional morphology. It is clear that PyC and SiC deposit alternately on the surfaces of each other as designed, which display finely multilayered structure.

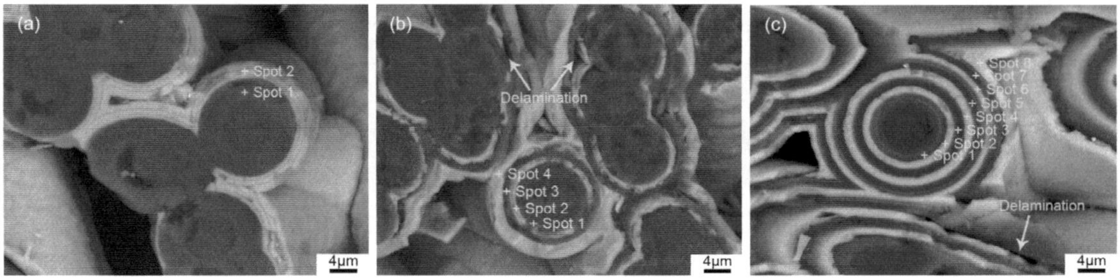


**Fig. 1 BSE images of polished cross-sectional morphology of CS1 (a), CS2 (b) and CS4 (c).**

As shown in Fig. 2, CTEs of the three composites gently increase from 800 to 1200 °C and then decrease sharply reaching the minimums at around 1500 °C. Beyond 1500 °C, CTEs increase to peak values at about 1800 °C and fall to valley values at about 2100 °C, then again increase linearly.

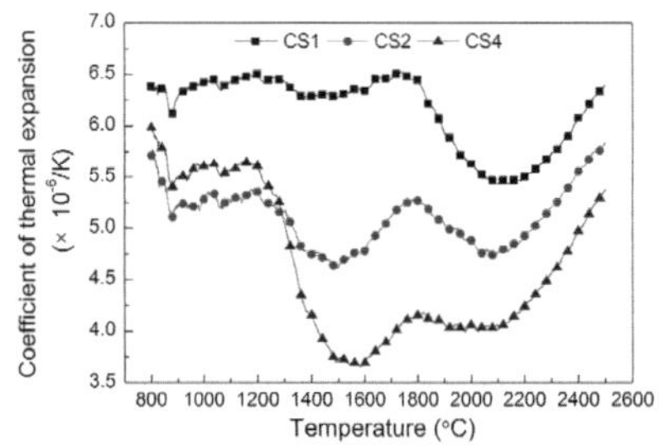


**Fig. 2 CTE curves of C/(PyC–SiC)$_n$ composites from 800 to 2500 °C.**

From Fig. 3(a), the specific heat of C/(PyC–SiC)$_n$ composites increase exponentially with the increase of temperature. The thermal diffusivity can be expressed by parabolic curves (Fig. 3(b)). Thus, the dependence of thermal conductivity with temperature can be obtained (Fig. 3(c)).

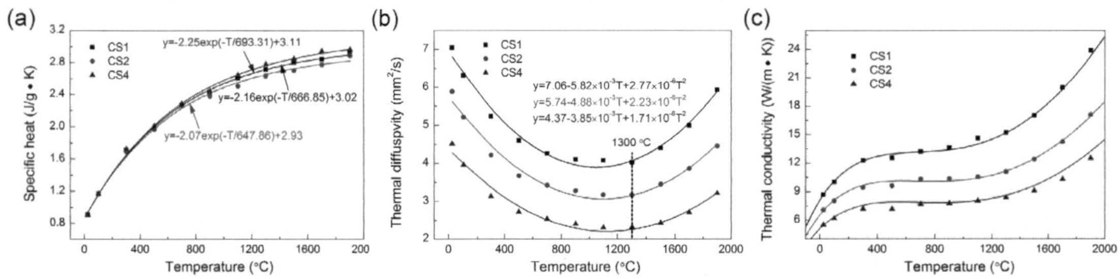


**Fig. 3 Specific heat (a), thermal diffusivity (b) and thermal conductivity (c) of C/(PyC–SiC)$_n$ composites.**

## CONCLUSIONS

With increasing the sequences number (n), the CTEs of the composites decline owing to the increase of interfacial delamination, providing room for the thermal expansion. The thermal diffusivity and thermal conductivity also decrease due to the enhancement of phonon-interface scattering resulted from the increasing number of interfaces.

## REFERENCES

1. Wang, J.P., Lin, M., Xu, Z., Zhang, Y.H., Shi, Z.Q., Qian, et al. (2009). "Microstructure and mechanical properties of C/C–SiC composites fabricated by a rapid processing method", *J. Eur. Ceram. Soc.*, Vol. 29(14), p.3091-3097.

2. Yuan, G.M., Li, X.K., Dong, Z.J., Xiong, X.Q., Rand, B., Cui, Z., et al. (2014). "Pitch-based ribbon-shaped carbon-fiber-reinforced one-dimensional carbon/carbon composites with ultrahigh thermal conductivity", *Carbon*, Vol. 68, p.413-425.

3. Jia, Y., Li, K.Z., Xue, L.Z., Ren, J.J., Zhang, S.Y., Li, H.J. (2017). "Mechanical and electromagnetic shielding performance of carbon fiber reinforced multilayered $(PyC–SiC)_n$ matrix composites", *Carbon,* Vol. 111, p.299-308.

# TiCo/Graphene based nanocomposites and their activity towards oxygen reduction reaction

Luque-Centeno, J. M.[1], Martínez-Huerta, M.V.[2], Lemes, G.[1], Sebastián, D.[1], Pastor, E.[3], Lázaro, M.J.[1*]

[1]Instituto de Carboquímica (CSIC), Miguel Luesma Castán 4, 50018 Zaragoza, Spain.
[2]Instituto de Catálisis y Petroleoquímica (CSIC), Marie Curie 2, 28049 Madrid, Spain.
[3]Instituto de Materiales y Nanotecnología, Dpto. Química-UD Química-Física, Universidad de La Laguna, Apartado Postal 456, La Laguna (Tenerife), Spain.
*mlazaro@icb.csic.es

## INTRODUCTION

Alkaline electrolyte based fuel cells appear as a good power source to solve some of the problems related to proton exchange membrane ones. The use of an anionic electrolyte allows the use of cost-effective non-precious metals as catalysts, due to improved kinetics at high pH, and reduces corrosion phenomena related to precious metals.[1]

Titanium-based materials such as nitrides, carbides and carbonitrides have exhibited good properties as Pt support to decrease the noble metal loading with a high stability.[2,3] On the other hand, cobalt has been widely reported in literature as a good catalyst for the oxygen reduction (ORR) and oxygen evolution (OER) reactions in alkaline medium.

The aim of this work is the investigation of composites based on graphene with titanium and cobalt as non-precious metals for the ORR and the OER in alkaline medium. Reduced graphene oxide (rGO) has been used as carbon support due to its great physico-chemical properties like high surface area, good chemical and electrochemical resistance and high electrical conductivity.

## MATERIALS AND METHODS

TiCo/rGO composites were synthesized by a sol-gel methodology, followed by thermal treatment under nitrogen atmosphere at 700 and 800 ºC. Ti/rGO and Co/rGO were prepared in the same conditions at 800 ºC. The as-synthetized composites were physico-chemically characterized by X-Ray Diffraction (XRD), Elemental Analysis (EA), and Inductively Coupled Plasma - Mass Spectrometry (ICP-MS).

The electrocatalytic activity and stability towards ORR and OER have been determined in a three-electrode system, using a glassy carbon Rotating Disk Electrode (RDE) in alkaline medium (0.1M NaOH).

## RESULTS AND DISCUSSION

The rGO-supported TiCo-based materials show a metal content (Ti/Co) in the range 12-27 wt%, as reported in Table 1. The nitrogen content is higher in the samples containing Co, with up to 8 wt% for the sample treated without Ti. TiCo/rGO-800 presents also lower Ti % than the sample TiCo/rGO-700.

| Sample | C(%) | H(%) | N(%) | Ti(%) | Co(%) |
|---|---|---|---|---|---|
| Ti/rGO | 53.8 | 2.7 | 2.3 | 22.4 | - |

| | | | | | |
|---|---|---|---|---|---|
| Co/rGO | 59.5 | 0.4 | 8.2 | - | 24.7 |
| TiCo/rGO-700 | 20.7 | 0.3 | 2.0 | 23.8 | 23.5 |
| TiCo/rGO-800 | 36.7 | 0.7 | 4.2 | 12 | 26.7 |

**Table 1.** Chemical composition (wt%) of the rGO-supported composites (EA, ICP-MS).

XRD was performed to investigate the crystal phases of the composites (Figure 1A). In all cases, a peak at a 2□ of 25° assigned to diffraction of graphite carbon (002) is observed. The diffraction peaks characteristic of $TiO_2$ anatase and rutile phases were also observed in the composites with Ti. The most remarkable difference between composites is the appearance of $TiCoO_3$ alloy phase for the sample treated at 700 °C.

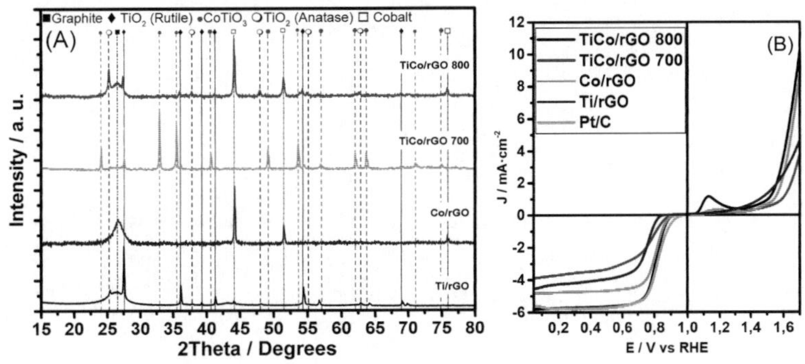


**Fig 1.** A) XRD patterns of TiCo/rGO, Co/rGO and Ti/rGO composites. B) Activity of composites towards ORR and OER.

The electrocatalytic activity towards ORR is enhanced for the catalysts treated at 800 °C (Figure 1B). This could be related to the presence of non-alloyed Co nanoparticles and the higher percentage of nitrogen which may act as catalyst for these reactions. Similar trends were observed in OER, being TiCo/rGO-800 the most active catalyst. An oxidation peak at 1.15 V vs. RHE should be related to oxidation of Co nanoparticles.

## CONCLUSIONS

Sol-gel process is presented as a good method for the synthesis of $TiO_2$/rGO composites modified with cobalt. The graphene-based bimetallic TiCo composites show promising properties as electrocatalysts toward ORR and OER.

## ACKNOWLEDGMENTS

Authors acknowledge financial support given by Spanish Ministry of Economy and Competitiveness (MINECO) through projects ENE2014-52158-C2-1-R and 2-R (co-founded by FEDER). J. M. Luque and G. Lemes also thank MINECO and Aragon Government, respectively, for their Ph.D. grants.

## REFERENCES

1. Y. Wang, D.Y.C. Leung, J. Xuan, H. Wang, A review on unitized regenerative fuel cell technologies, part-A: Unitized regenerative proton exchange membrane fuel cells, Renewable and Sustainable Energy Reviews, 65 (2016) 961-977.

2. M. Roca-Ayats, E. Herreros, G. García, M.A. Peña, M.V. Martínez-Huerta, Promotion of oxygen reduction and water oxidation at Pt-based electrocatalysts by titanium carbonitride, Applied Catalysis B: Environmental, 183 (2016) 53-60.
3. M. Roca-Ayats, G. García, J.L. Galante, M.A. Peña, M.V. Martínez-Huerta, TiC, TiCN, and TiN Supported Pt Electrocatalysts for CO and Methanol Oxidation in Acidic and Alkaline Media, The Journal of Physical Chemistry C, 117 (2013) 20769-20777.

# Transverse Structure Control of Mesophase Pitch Based Carbon Fiber

Sun Zhulin[1], Yang Changlin[2], Lu Yonggen[1*]

[1]College of Materials Science and Engineering, Donghua University, Shanghai 201620, China
[2]College of Chemistry, Chemical Engineering and Biotechnology, Donghua University, Shanghai 201620, China
*yglu@dhu.edu.com

## INTRODUCTION

Mesophase pitch based carbon fibers (MPCFs) usually crack due to radiation pattern on the transverse section, which can be modified by blending specific additives in the mesophase pitch[1,2]. This work is aimed at alleviating this structure by doping polyacrylonitrile (PAN) into mesophase pitch. The modified MPCFS were prepared through spinning, stabilization, and carbonization processes. It is found that very small amount of PAN can not only alleviate the carbon fiber cracking by modifying its transverse structure, but also exert some co-carbonization effect which contributes to increasing its crystalline sizes.

## MATERIALS AND METHODS

### Material

A commercial naphthalene mesophase pitch was obtained from Changzhou Herman Carbon Company. The PAN was supplied by Shanghai Petrochemical Corp., which was synthesized by water-phase precipitation copolymerization through acrylonitrile, methyl acrylate and itaconic acid.

### Methods

*Mixing*. A predetermined amount of PAN copolymer was dissolved in DMSO in a glass beaker. Subsequently, a certain amount of the mesophase pitch powder was added into the solution to form a paste. The paste was milled in a ball-milling machine for enough long time to obtain a homogeneous mixture, and then the mixture was dried at 80 °C in vacuum. In order to make the doped PAN molecules cyclize, as well as to remove the residual DMSO, the mixture was heated at 300 °C for 6 h in vacuum.

*Preparation of the* MPCFs. Pitch fibers were prepared by melt-spinning at an optimum temperature on a laboratory monofilament spinning apparatus. The spinneret used has a round nozzle of 0.28 mm in diameter and 3.0 in ratio of length to diameter. The pressure was maintained at 0.5 MPa during spinning. The derived pitch fibers were stabilized in an oven in air by a procedure following: at 5

°C/min from room temperature to 150 °C, and then at 0.5 °C/min to 270 °C. Carbonization was conducted in an intermittent carbonization furnace in nitrogen, and the heating rates of the experiment below 300 , 300~650 and 650~1300 °C were 5, 2 and 10 °C/min, respectively.

## RESULTS AND DISCUSSION

Field-emission scanning electron microscopy (SEM, SU8010) with energy dispersive spectroscopy (EDS), was used to observe the transverse section of mesophase pitch fibers and the dispersity of PAN.

**Fig. 1.** shows the distribution of nitrogen in pitch fiber. It is noted that the dots in (b) standing for nitrogen generating signal are uniformly distributed. Since the N was introduced by PAN, it is considered the PAN molecules were distributed homogeneously.

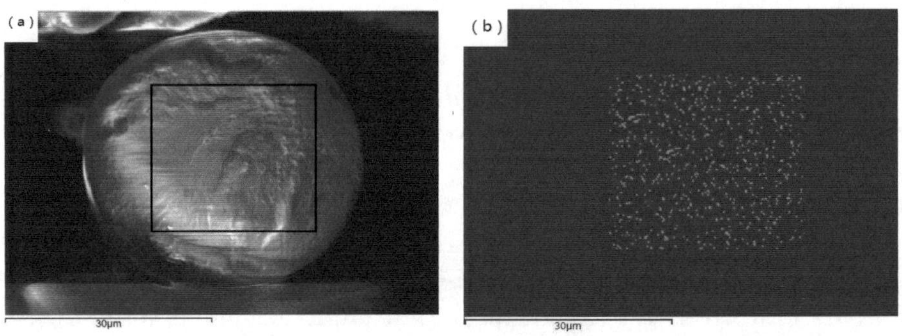


Fig. 2 Mapping images of mesophase pitch fiber
(a) SEM image of pitch fiber doped with 0.3 wt. % of PAN, (b) Nitrogen elemental mapping image of the square area in (a).

**Fig. 2.** shows the transverse section micrographs of the unmodified and the 0.3 wt. % PAN doped MPCFs. It's noted that the later one exhibited a greater folded structure and the crack along the fiber axis was suppressed.

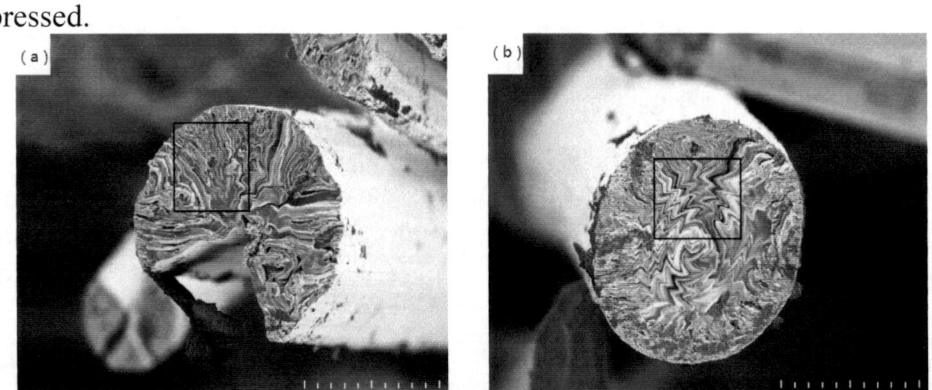


Fig. 2. Transverse section structures of MPCFs

**1.** Unmodified, (b) doped with 0.3 wt. % of PAN

## CONCLUSIONS

In the present work, mesophase pitch doped with 0.1 to 0.5 wt. % PAN was successfully spun into fibers. The spinnability is better when PAN content is below 0.3 wt. %. After carbonization, the carbon fibers doped with 0.3wt. % PAN exhibited a greater folded structure and the crack along the fiber axis was suppressed. Meanwhile the PAN promoted the graphitic structure formation of the carbon fibers, which is reflected in its larger crystalline sizes. The La and Lc values of MPCFs doped with 0.3wt. % PAN increased by 38.36 % and 16.73 %, respectively, compared with the undoped one .

## REFERENCES

1. Cho, T., Lee, Y. S., Rao, R., Rao, A. M., Edie, D. D., Ogale, A. A. (2003). "Structure of carbon fiber obtained from nanotube-reinforced mesophase pitch", Carbon., Vol.41(7), p.1419-24
2. Rebecca, M. A., Anderson, D. P, Ogale, A. A. (2013). "Carbon black modification of mesophase pitch-based carbon fibers", Carbon., Vol.59,p.40-8

# Two-Dimensional Carbon-Encapsulated $Fe_3O_4$ Nanosheets Grown Via Oriented Attachment Mechanism

Jinyu Ning[1], Minhong Kong[1], Huaihe Song[1,2], Jisheng Zhou[1,2*]

[1] State Key Laboratory of Chemical Resource Engineering, Beijing Key Laboratory of Electrochemical Process and Technology for Materials, Beijing University of Chemical Technology, Beijing 100029, P. R. China
[2] Changzhou Institute of Advanced Materials, Beijing University of Chemical Technology, Jiangsu, P. R. China.
* zhoujs@mail.buct.edu.cn (Zhou, J)

## INTRODUCTION

Since the great progress in graphene [1], two-dimensional (2D) nanomaterials have been one of the most attractive research directions for numerous scientists. And the Oriented Attachment (OA) mechanism is one of the most promising approaches to controlling anisotropy growth of nanomaterials via irreversible attachment between nanocrystal surfaces.[2, 3] Moreover, most cases related with OA growth are carried out in the solution systems with the help of surfactant. Organic ligand molecules as the surfactant play an important role in the anisotropic growth of nanocrystals by coating selectively on the surface of primary nanocrystals. Therefore, attachment between the primary nanocrystals occurs on the crystal face without the surfactant. Interestingly, the carbon layers can also act similar functions as surfactants to guide the OA growth of nanocrystals. Thus, novel carbon encapsulated nanocrystals with anisotropic structures can be tailored, which largely expands OA growth in the carbon coated materials.

## MATERIALS AND METHODS

The products, $Fe_3O_4$@C, were prepared by the *in situ co-pyrolysis* method, which is an effective and convenient technology to produce carbon nanomaterials. In our previous work, we have employed *in situ co-pyrolysis* method to synthesize various nanomaterials. [4, 5] And here are the main preparation processes. Firstly, Ferrocene (0.3 g) with distilled water (58 μL, ca. 0.058 g) was added into a 30 ml autoclave. Then, the sealed autoclave was put into muffle furnace and heated from the room temperature to 500 °C with a heating rate of 10 °C/min. After heating at 500 °C for 2h, the autoclave was cooled naturally to the room temperature. The final products, 2D $Fe_3O_4$@C nanosheets, are black powders, which were obtained by washing 3-4 times using alcohol and subsequently drying overnight at vacuum oven at 60 °C.

## RESULTS AND DISCUSSION

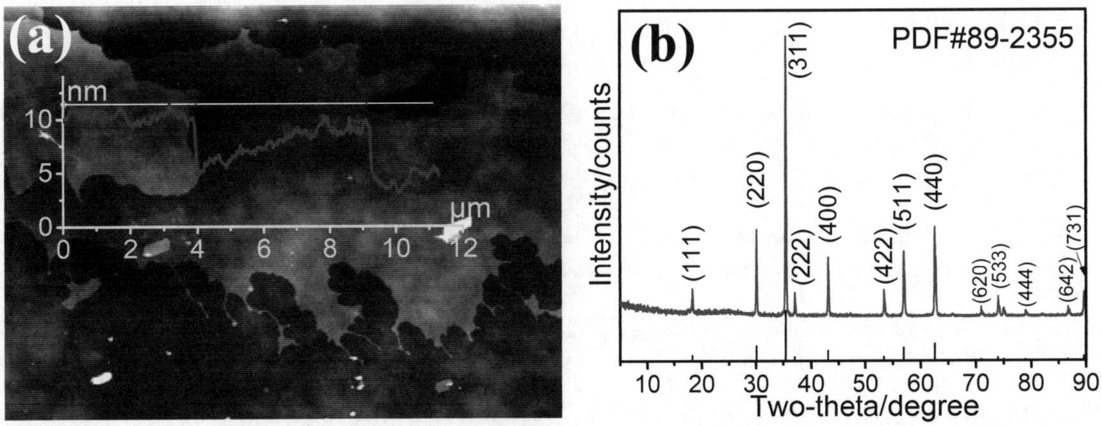


Figure 1. (a) The AFM image of the $Fe_3O_4$@C and the measure results of the nanosheets' thickness, (b) the XRD patterns of the $Fe_3O_4$@C nanosheets.

The AFM figures of the products are shown in **Figure 1a** where the surface features have been clearly revealed. It is obvious that the surface is very flat and thin and the thickness is ca. 5 nm. More interestingly, some extend beard-like nanostructures and little nanosheets show on the fringe of the $Fe_3O_4$@C nanosheets that should be the trace relating the OA growth mechanism. The XRD patterns (**Figure 1b**) reveal that the final products can be well indexed to a pure $Fe_3O_4$ phase according to the JCPDS card (PDF#89-2355). And the diffraction peaks without any other additional peaks are very sharp, indicating the high crystalline degree of the $Fe_3O_4$.

**Figure 2** demonstrates the rate capability of the samples at various current densities for LIBs. The first discharge specific capacities of the $Fe_3O_4$@C nanosheets is 1250 mA h $g^{-1}$ at 50 mA $g^{-1}$ and the average specific capacities can be up to 740, 703, 587, 470, 394, 335 and 263 mA h $g^{-1}$ at 50, 100, 200, 500, 1000, 2000 and 5000 mA $g^{-1}$, respectively. When the current density returns back to 50 mA $g^{-1}$, the average specific capacity of the $Fe_3O_4$@C nanosheets can recover to 690 mA h $g^{-1}$.

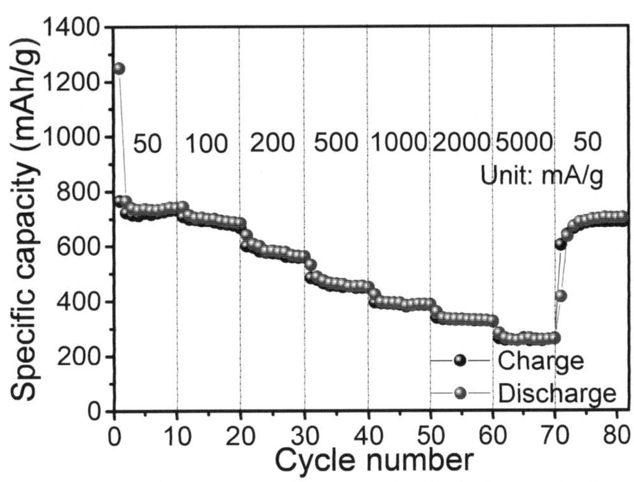


Figure 2. The rate performance of the $Fe_3O_4$@C electrode for the LIBs.

## CONCLUSIONS

The novel $Fe_3O_4$@C nanosheets were prepared by a facile one pot *in situ co-pyrolysis* method. The formation mechanism of 2D $Fe_3O_4$@C should be attributed to the oriented attachment mechanism which can create diverse anisotropic nanomaterials. Furthermore, as anode materials for LIBs, $Fe_3O_4$@C nanosheets exhibit excellent rate capability, which should be attributed to the carbon coated 2D nanostructures.

# REFERENCES


1. Geim, A. K. and Novoselov, K. S. (2007). "The rise of graphene", *Nat. Mater.* , Vol.6(3), p.183-191.

2. Liang, H., Zhao, H., Rossouw, D., et al. (2012). "Silver Nanorice Structures: Oriented Attachment-Dominated Growth, High Environmental Sensitivity, and Real-Space Visualization of Multipolar Resonances", *Chem. Mater.* , Vol.24(12), p.2339-2346.

3. Yu, B. B., Zhang, X., Jiang, Y., et al. (2015). "Solvent-induced oriented attachment growth of air-stable phase-pure pyrite FeS2 nanocrystals", *J. Am. Chem. Soc.* , Vol.137(6), p.2211-2214.

4. Zhou, J., Song, H., Chen, X., et al. (2009). "Oxidation Conversion of Carbon-Encapsulated Metal Nanoparticles to Hollow Nanoparticles", *Chem. Mater.* , Vol.21(15), p.3730-3737.

5. Wu, B., Song, H., Zhou, J., et al. (2011). "Formation mechanism of carbon-encapsulated iron nanorods in a co-carbonization process", *Carbon*, Vol.49(3), p.890-894.

# Water desalination with graphene oxide/graphene based membranes

Aaron Morelos-Gomez,[1*] Rodolfo Cruz-Silva,[1,2] Hiroyuki Muramatsu,[2] Josue Ortiz-Medina,[1] Takumi Araki,[1,3] Tomoyuki Fukuyo,[4] Syogo Tejima,[1,3] Kenji Takeuchi,[1,2] Takuya Hayashi,[1,2] Mauricio Terrones,[2,5] Morinobu Endo[1,2*]

[1]Global Aqua Innovation Center, Shinshu University, 4-17-1 Wakasato, Nagano 380-8553, Japan

[2]Institute of Carbon Science and Technology, Shinshu University; 4-17-1 Wakasato, Nagano 380-8553, Japan.

[3]Research Organization for Information Science & Technology, 2-32-3, Kitashinagawa, Shinagawa-ku, Tokyo, 140-0001, Japan

[4] Showa Denko K.K., Institute for Advanced and Core Technology, 1-1-1, Ohnodai, Midori-ku, Chiba-shi, Chiba, 267-0056, Japan

[5] Department of Physics, Department of Chemistry, and Department of Materials Science and Engineering, The Pennsylvania State University; University Park, Pennsylvania 16802, USA

* amorelos@shinshu-u.ac.jp, endo@endomoribu.shinshu-u.ac.jp

## INTRODUCTION

Graphene oxide (GO) membranes for nanofiltration have been studied recently with room for improvement in salt rejection, cross-flow shear resistance, scalability, chlorine resistance, among other themes.[1)] Previous reports have shown GO membranes operating in static cells, dead end flow and cross flow systems. The latter, is widely used for large scale clean water output; based on applying high pressure (5 MPa) with a water feed flowing on the membrane surface. The majority of the reported GO membranes rely on dead end flow measurements and very few on cross flow, due to the low water shear resistance.[2)] For cross flow shear stability, carbon nanotubes (CNTs) have been introduced in the GO membrane,[2)] however CNTs are much larger than the size of hydrated ions therefore high salt rejection is not expected. For high salt rejection, vertically aligned GO has achieved salt rejection above 90%, however for a static cell due to the extremely low mechanical stability under pressure and difficulty in membrane preparation.[3)] Therefore, an easy and scalable method for graphene preparation is highly desirable, for this spray-coating could be feasible.

Here, a new type of hybrid layered membranes, made by simple spray-coating of GO/few-layer-graphene (FLG)/surfactant has been studied as desalination membranes. These membranes were deposited onto commercial polysulfone (PSU) support membranes modified with an adhesive polymer. Followed by thermal treatment and crosslinking to improve the robustness of the membrane. This membrane exhibits excellent resistance to high water shear, good salt rejection, chlorine resistance and long term operation stability.

## MATERIALS AND METHODS

### Material

GO was prepared following the method reported by a modified Hummers method. Graphite was mechanically exfoliated with an ultrasonic tip homogenizer in water using a surfactant as stabilizer during 20 h at 10 $^{o}$C. Afterwards, the dispersion was centrifuged discarding the precipate. The remaining dispersion consists of few-layer-graphene sheets dispersed in water (FLG). The mixtures of GO slurry and FLG dispersion were done at different mass ratios. A composite membrane was prepared by soaking a polysulfone membrane in an adhesive polymer solution. Later a dispersion of GO/FLG was spray coated onto the polysulfone support using an airbrush. Afterwards, the membrane was dried in an oven at 100 $^{o}$C for 1 h and left to cool to room temperature. Crosslinking was carried out by submerging the membrane in a divalent ion solution in ethanol/water (1:3). Excess crosslinker was removed by washing.

## RESULTS AND DISCUSSION

From scanning electron microscopy (SEM) observations the GO sheets exhibited an average length of 12 μm and the FLG structures exhibited two length distributions at 0.5 μm and 3 μm. In the spray coated FLG a smooth surface was observed and in the GO/FLG membranes top view, the FLG appeared between GO sheets, and some GO wrinkles could be observed. The thicknesses of the membranes were in the range 26 – 33 nm. Transmission electron microscopy (TEM) images of the FLG sheets revealed that each one has between 2 and ca. 16 layers of graphene, and are partially covered by surfactant, with no observed pores and the GO sheets exhibited few pores with sizes lower than 2 nm that cover 1% of the GO sheet area.

The membranes were tested under cross-flow velocities between 200 and 1000 ml/min at 1 MPa for 1 hour during each velocity. During all the tests optical microscope images showed no evidence of peel off of the membrane made with only GO, indicating excellent water shear resistance. Salt rejection for GO/FLG membranes were measured at pressures between 1 and 5 MPa for 1 day at each pressure. The water permeation exhibits a linear behaviour against pressure, indicating that the membranes were stable during the measurements. For GO content above 45% the water permeation was similar and for lower content is decreased due to the size of FLG and the present surfactant. Similar behaviour was observed for NaCl rejection with values between 80% and 90%, and decreasing for GO content below 45% (Figure 1). Each membrane was operated for a total of 5 days varying pressure with constant cross flow velocity and stable desalination performance, demonstrating the possibility for long term operation.

In addition, salt rejections with different ions and concentrations demonstrate that the Gibbs-Donnan exclusion mechanism is responsible for the desalination performance. With higher ionic strength of the feed solution there is a decrease in salt rejection. Furthermore, the membrane could separate a mixture of anionic (Acid Blue 9) and cationic (Rhodamine B) dyes favouring the permeation of cationic dyes. Initially the mixture had 50% wt. anionic and cationic dyes, later the permeate water had up to 87% wt. of Rhodamine B.

Chlorine treatment is widely used in desalination plants to take care of bio- and organic fouling. Biofouling is one of the greatest drawbacks in membrane technology since it could decrease the operation lifetime. Pure GO and GO/FLG membranes were kept in a solution of 200 ppm sodium hypochlorite for one day and their desalination performance was measured with and without chlorine treatment. Pure GO membranes exhibited a decrease in NaCl rejection from 85% to 75% and GO/FLG from 83% to 81%, and both membranes increased their water permeation by four times. This finding shows that GO/FLG has a much better chlorine resistance than pure GO due to FLG.

## CONCLUSIONS

The spray-coating deposition is an easy, scalable and environmental friendly method to prepare GO membranes. This membrane could withstand high cross flow velocities and lower velocities for several days with stable performance due to the excellent adhesion between GO and the substrate provided by the polymeric adhesive. Good salt rejection was achieved by both pure GO and GO/FLG, with a wide range of GO contents (> 45%). The Gibbs-Donnan exclusion could be given by the functional groups of GO, the surfactant and divalent ions (used for crosslinking). When FLG was introduced high chlorine resistance could be achieved due to the low reactivity of FLG. These membranes exhibit excellent resistance to high water shear, good salt rejection, chlorine resistance and long term operation stability. Therefore they are very promising for future research and applications. They can be used for bioreactors, protein separation, pharmacy, food industry, wastewater treatment, among other possibilities.

## REFERENCES

1. Cornell, A. B., Emerald, C. D. and Hong, G.-D. (2011). "Effect of $CO_2$ on global climate change", *J. Env. Eng.*, Vol.11(8), p.21-24

1. Goh, K., Karahan, H. E., Wei, L., Bae, T., Fane, A. G., Wang, R. et al. (2016). "Carbon nanomaterials for advancing separation membranes□: A strategic perspective", *Carbon*, Vol. 109, p. 694-710

2. Goh, K., Jiang, W., Karahan, H. E., Zhai, S., Wei, L., Yu, D., et al. (2015). "All-Carbon Nanoarchitectures as High-Performance Separation Membranes with Superior Stability", *Adv. Funct. Mater.*, Vol. 25, 7348–59.

3. Abraham, J., Vasu, K. S., Williams, C. D., Gopinadhan, K., Su, Y., Cherian, C. T., et al. (2017). "Tunable sieving of ions using graphene oxide membranes", *Nat. Nanotechnol.*, doi:10.1038/nnano.2017.21.

# AUTHOR INDEX


Alegre, C. ....60
Alegre, Cinthia ....68
Alonso-Morales, Noelia ....125
Arai, Masahiro ....141
Araki, Takumi ....215
Areerob, Yonrapach ....111
Arenillas, A. ....93
Astromskas, Gvidas ....170
Bahn, Emanuel ....152
Bakos, István ....131
Barkauskas, Jurgis ....158, 170
Barrios-Yáñez, G. ....113
Behnisch, T. ....101
Bello, A. ....184
Berke, Barbara ....27, 191
Bertóti, Imre ....131, 191
Bing, Cao ....9
Blume, Raoul ....33
Bohm, R. ....101, 160
Boonyoung, Pawan ....6
Bukauskas, Virginijus ....170
Calvo, Luisa ....125
Canal-Rodriguez, M. ....93
Cao, Bin ....30
Cao, Bing ....81
Chabal, Yves ....33
Chaiwat, Weerawut ....47
Charinpanitkul, Tawatchai ....47, 144
Chartrand, Patrice ....200
Cházaro-Ruiz, L. F. ....15, 113
Chen, X. ....24, 35
Chen, Xiaohong ....19, 30, 63, 128, 136, 150, 187
Chen, Yaxin ....19
Cherif, C. ....160
Cho, Donghwan ....162
Cho, Kwang-Youn ....193
Choi, Hoikil ....96
Chung, D. D. L. ....58
Cortés-López, Alejandro J. ....175
Crean, Carol ....198
Cruz-Silva, Rodolfo ....215
Czakkel, Orsolya ....27, 152
Dangbegnon, J. K. ....184
Ding, Yuanfei ....122
Dong, Yanfeng ....138
Dong, Yue ....63
Dreyer, Kathleen ....33
Eiad-Ua, Apiluck ....13
Elías, Ana Laura ....147
Endo, Morinobu ....215
Eto, Makoto ....50
Fajardo-Diaz, Juan L. ....175
Faungnawakij, Kajornsak ....13
Fouquet, Peter ....152
Fuentes, Erika ....33
Fukuyo, Tomoyuki ....215
Gaidukevic, Justina ....158, 170
Gálvez, María E. ....68
Gao, Feng ....41
García, Gonzalo ....195
Geng, Chuang ....41
Genys, Povilas ....158
Gheribi, Aïmen E. ....200
Gilarranz, Miguel A. ....125, 147
González-Castillo, A. C. ....113
Goto, Yuko ....179
Gude, M. ....101, 160
Gunpum, Wachiraporn ....13
Guo, Ling-Jun ....77
Guo, Shasha ....19
Guo, X. ....24
Gutiérrez-Martínez, J. J. ....104
Hansen, Thomas ....33
Hashimoto, Ryo ....141
Hayashi, Takuya ....172, 215
Heras, Francisco ....125, 147
Hibbitts, David ....33
Hsieh, Shu-Huei ....189
Hu, Lanqing ....108
Hu, Mingxiang ....53
Huang, Fei ....3
Hwang, Daekyun ....162
Inada, Ryoji ....119
Irisawa, Toshihira ....141
Ishii, Takafumi ....85, 106
Iwashita, Norio ....21
Jager, H. ....160
Jan, Ya-Wun ....189
Jaroniec, Mietek ....56
Jia, Dianzeng ....63
Jia, Husheng ....122
Jia, Mengqiu ....150
Jia, Yan ....203
Kado, Yuya ....44
Kaneko, K. ....1
Kaneko, Katsumi ....172
Kang, F. ....116
Kang, Feiyu ....53
Kannari, Naokatsu ....85, 106
Kerdnawee, Konrat ....144
Khaleed, A. A. ....184
Kim, Woo-Sik ....193
Kinumoto, Taro ....50, 75
Kirsten, M. ....160
Kodama, Masaya ....90
Kong, Minhong ....128, 212
Kong-Gij, Bunsita ....6
Ku, Chong Rae Park Bon-Cheol ....96
Kubo, Shingo ....134
Kunapisitkul, Sataporn ....6

# AUTHOR INDEX


László, Krisztina ....27, 131, 152, 191
Lazaro, M. J. ....60, 68, 87, 195, 206
Lee, Jae Kwan ....96
Lei, Yu ....147
Lemes, G. ....87, 206
Li, B. ....116
Li, Dan ....187
Li, Kezhi ....203
Li, Q. ....116
Li, Rui ....138
Liang, G. ....116
Liang, J. ....71, 155
Liu, Donghai ....17
Liu, J. ....71, 155
Liu, M. ....116
Liu, Sitong ....136
Liu, Weifeng ....108
Liu, Xuguang ....108, 122
Loffler, M. ....160
López, D. Domene ....38
López-Urías, Florentino ....175
Lu, Yonggen ....3, 209
Luchnicov, Valeriy ....164
Luque-Centeno, J. M. ....87, 206
Lv, Ruitao ....53
Lv, Wei ....17
Ma, Y. ....71
Ma, Zhaokun ....19, 30, 63, 79
Maie, Takuya ....85, 106
Mališauskaite, Ieva ....170
Manyala, N. ....184
Martinez-Huerta, M. V. ....60, 87, 206
Martin-Gullón, I. ....38, 93
Matos, J. ....113
Matsumura, Kazuki ....75
Matsuoka, Miki ....50
Matulaitiene, Ieva ....170
Menakanist, Karanick ....144
Menéndez, J. A ....93
Meng, Qinghan ....81
Meng, Yang ....9
Miao, Yanqin ....122
Mikoliunaite, Lina ....158
Mohai, Miklós ....131, 191
Montero, César ....195
Morelos-Gomez, Aaron ....215
Morkvenaite–Vilkonciene, Inga ....158
Muhle, U. ....160
Muñoz-Sandoval, Emilio ....175
Muramatsu, Hiroyuki ....215
Mwakikunga, B. W. ....184
Nagy, Balázs ....131, 152
Nakada, Masaru ....168
Nariduka, Kumi ....85, 106
Nguyen, Dinh Cung Tien ....193
Niaura, Gediminas ....170
Nieto-Monge, M. J. ....60
Ning, Jinyu ....212
Nishi, Masayasu ....65
Nueangnoraj, Khanin ....6
Oh, Won-Chun ....111, 166, 193
Okuda, Haruki ....168
Ono, Kiminori ....168
Ono, Kohei ....50
Opasanon, Napon ....144
Ortiz-Medina, Josue ....215
Osawa, Eiji ....172
Ouzilleau, Philippe ....200
Ozaki, Jun-Ichi ....85, 106
Paklonskaite, Ieva ....158
Park, Ok-Kyung ....96
Pastor, E. ....60, 87, 206
Pastor, Elena ....195
Pauliukaite, Rasa ....170
Peng, Jian ....77
Peng, Xiyue ....41
Pérez-Galera, J. ....38
Petroniene, Jurate Jolanta ....158
Piña-Salazar, Elda-Zoraida ....172
Porcar, Lionel ....27
Qin, Lei ....108
Qin, X. ....116
Qinghan, Meng ....9
Qiu, Jieshan ....41, 138
Qu, Jiangying ....41
Ramanavicius, Arunas ....158
Ramos-Fernández, G. ....38, 93
Rangel-Mendez, J. R. ....104, 113
Rangel-Méndez, René ....15
Rao, Radhika ....33
Rellinghaus, B. ....160
Rivera, Luis M. ....195
Rodríguez, Juan J. ....125
Ruiz-Cornejo, J. C. ....60
Ruiz-García, Cristina ....125, 147
Saito, Yukie ....164, 179
Sakai, Toshio ....172
Sakurai, Yoji ....119
Sano, Noriaki ....144
Santoyo-Cisneros, Rigoberto ....15
Sato, Satoshi ....99, 134
Schlögl, Robert ....33
Sebastian, D. ....60, 68, 87, 206
Slade, Robert ....198
Smith, Emily A M ....198
Soneda, Yasushi ....21, 44, 90
Song, H. ....24, 35
Song, Huaihe ....19, 30, 63, 79, 128, 136, 150, 187, 212
Spacie, Chris ....198
Srisuma, Prakitr ....144

# AUTHOR INDEX


Stirling, Chris ........ 198
Sun, Zhulin ........ 209
Suttiponparnit, Komkrit ........ 47
Suwattanapongtada, Nuttawat ........ 144
Taira, Saori ........ 65
Takagi, Hideyuki ........ 65
Takasu, Koji ........ 85
Takeuchi, Kenji ........ 215
Takigami, Machiko ........ 85, 106
Tamon, Hajime ........ 144
Tanabe, Yasuhiro ........ 73, 141
Tanaka, Fumihiko ........ 168
Tanaka, Haruki ........ 134
Tascón, Juan M. D. ........ 152
Tejima, Syogo ........ 215
Termvidchakorn, Chompoopich ........ 144
Terrones, Mauricio ........ 53, 147, 215
Tessonnier, Jean-Philippe ........ 33
Thieme, M. ........ 101, 160
Thonganatakul, Oukrit ........ 47
Togo, Takaaki ........ 99
Tojo, Tomohiro ........ 119
Iirisawa, Toshihira ........ 73
Toyoda, Masahiro ........ 50, 75
Treeweranuwat, Panudetch ........ 6
Trusovas, Romualdas ........ 170
Tsuchiya, Mitsuki ........ 182
Tsumura, Tomoki ........ 50
Tsumuta, Tomoki ........ 75
Tzeng, Yi-Ren ........ 189
Uda, Michimasa ........ 21
Uematsu, Katsumasa ........ 73
Umeda, Daichi ........ 99
Urita, Koki ........ 172
Vaewhongs, Parintorn ........ 47
Villar-Rodil, Silvia ........ 152
Viriya-Empikul, Nawin ........ 13
Wadsungnoen, Pichaya ........ 6
Wang, Dingling ........ 79
Wang, X. ........ 71, 155
Wang, Xuzhen ........ 138
Watson, David ........ 198
Wolz, D. S. ........ 160
Xie, Yang'En ........ 79
Xu, Tao ........ 81
Xue, Y. ........ 71, 155
Yamada, Yasuhiro ........ 99, 134
Yamamoto, Tetsuya ........ 73
Yamauchi, Hiroshi ........ 21
Yanagisawa, Kenji ........ 182
Yang, Changlin ........ 209
Yang, Lan ........ 138
Yang, Quan-Hong ........ 17
Yang, X. ........ 155
Yang, Yongzhen ........ 108, 122
Yoshizawa, Noriko ........ 90
Yu, Jaesang ........ 96
Yu, Yunhua ........ 81
Yunhua, Yu ........ 9
Zhang, Chen ........ 17
Zhang, G. ........ 35
Zhang, Su ........ 63
Zhao, Weizhe ........ 3
Zheng, Jingxia ........ 122
Zhi, Lijie ........ 17
Zhi, Wanyuan ........ 81
Zhou, Chunli ........ 150
Zhou, J. ........ 24, 35
Zhou, Jisheng ........ 19, 30, 63, 128, 136, 187, 212
Zhou, Liangxiao ........ 3
Zschech, E. ........ 160